PHARMACEUTICAL ENGINEERING CHANGE CONTROL

SECOND EDITION

PHARMACEUTICAL ENGINEERING CHANGE CONTROL

SECOND EDITION

EDITOR
SIMON G. TURNER

CRC Press
Taylor & Francis Group
Boca Raton London New York

CRC Press is an imprint of the
Taylor & Francis Group, an **informa** business

CRC Press
Taylor & Francis Group
6000 Broken Sound Parkway NW, Suite 300
Boca Raton, FL 33487-2742

First issued in paperback 2019

© 2004 by Taylor & Francis Group, LLC
CRC Press is an imprint of Taylor & Francis Group, an Informa business

No claim to original U.S. Government works

ISBN-13: 978-0-8493-2061-3 (hbk)
ISBN-13: 978-0-367-39474-5 (pbk)

A CIP record for this book is available from the British Library.

Library of Congress Cataloging-in-Publication Data available on application

Visit the Taylor & Francis Web site at
http://www.taylorandfrancis.com

and the CRC Press Web site at
http://www.crcpress.com

Dedication

To Caroline

Foreword

Manufacturing pharmaceuticals is one of the most regulated human endeavours. The reason is that patients rightly expect pharmaceutical products to be correctly labeled and to be safe, pure and effective. The same expectation applies to veterinary products. The manufacturers of pharmaceutical products have a legal responsibility to keep their operations under control and to control the quality of their products. Good control is also good business, as yields will be consistent, failures will be minimal, and facilities and processes will remain fit for purpose. Change control is about the managed change of personnel, procedures, processes, premises and product quality.

Most regulations have been imposed on the industry following incidents and accidents. Most significantly, there is a mandatory quality system used to control the manufacture of pharmaceutical products known as *current Good Manufacturing Practice (cGMP)*, and the prefix "c" reminds us that practices are constantly improving.

Change control is a critical part of cGMP and, like all aspects of quality management, should be continuously improved. Change control is defined as a formal system by which qualified representatives of appropriate disciplines review proposed or actual changes that might affect the validated status of facilities, systems, equipment and processes. Done properly, validation and qualification procedures operate alongside change management and represent a planned, documented and science-based approach to the introduction of new systems and changes to existing systems. The systems can be physical and procedural.

This unique book gives manufacturers the necessary insight to improve the management of change, and covers the technical and management skills that are required. The complexity of modern pharmaceutical facilities means that no single person can be sufficiently knowledgeable in all facets of the business, and this is exemplified in the range of multidisciplinary skills that have had to be marshalled to produce this comprehensive book. The authors of the chapters are established experts and the full range of topics and examples are covered in the text.

Having spent more than 30 years in many aspects of pharmaceutical manufacture, I can appreciate the importance and value of this book and I am proud to be associated with it. This book will benefit design, construction, validation and maintenance engineers; control system software and hardware specialists; development scientists; production managers, supervisors and operators; as well as quality assurance professionals. It enables us to appreciate the process of change control, and thus minimizes the number of occasions when there are losses in quality, time and opportunity.

Stephen P. Vranch
Jacobs Engineering U.K. Ltd.

Preface

This book is written specifically for the professional working in the pharmaceutical industry. It may be regarded, at least in part, as a complementary text to Gillett's *Hazard Study and Risk Assessment in the Pharmaceutical Industry*,[1] by project and production managers, engineers, risk managers and technical staff involved in pharmaceutical design and operations. Business decision makers, environmental specialists and, especially, safety staff will also find this book useful. This book may also be regarded as a supplementary text for managers in healthcare manufacturing who have knowledge of generic works on change control that do not address key pharmaceutical industry issues specifically. Such generic works include the excellent IChemE modifications training module[2] and the book by Sanders.[3]

Change control is all about making sure that developments do not compromise the business operation. It is important to realize that change control is a management function, and that managers must take responsibility for ensuring adequate controls are exercised. There is a brief anecdote (see 'Nobody's Job' in Chapter 7) to illustrate the inexcusability of management complacency. There has to be control of all changes concerned with: engineering of the plant, process operations, raw material specification, product specification and manning, to name a few areas of manufacturing. The importance of training all those involved in controlling and managing change cannot be overemphasized. Training design, maintenance, and operating staff to recognize a change requires address by a management of change control process. This is probably the most important part of management's responsibility because once the change has been identified as a problem, the battle has been largely won. Without project and field personnel acting as efficient sensors to changes or modifications, changes cannot even be identified, let alone controlled. Improving the company's corporate memory can go a long way to improving this process by ensuring employees have a continuous working knowledge of what problems have arisen in the past, within their own company and the industry. The topics that training should cover can be identified by studying the topics in each chapter of this book, but the most important chapter for training is probably illustrating how unchecked changes cause problems, by case histories. Managers should not underestimate the importance of minor changes, which might be referred to as minor modifications. These are seemingly trivial changes, the consequences of which can be grossly underestimated and lead to major problems for a company (see the Murphy's anecdote in Chapter 7). Minor changes are typically those cheap jobs or design adjustments that need no approval. Beware of them; they *are* changes just the same and have the potential to be a project's or a company's undoing in a field of important business operations.

Change control does not appear to be an area that is well supported by specific change control techniques, as such. Safety problems can be identified by the techniques outlined for hazard study in Gillett's book. Change control ought to form one part of an organization's overall management system and often will be substantially integrated into a safety management system, which then employs the appropriate techniques to control safety, health and environmental (SHE) hazards. This means that general safety issues seem to be supported by hazard study techniques, but there may be many *changes* not fully covered by applying hazard study alone. We need additional change control procedures, to address this difference and target key areas for attention. There are recognized stages of operations, which can be checked, in line with your change control procedures, for important impacts on business and SHE. That is where this book can help you. Change control is an organization's prepared procedures "lying in wait," ready to be called into play on demand, to initiate the necessary response or technique. These procedures must be flexible enough to be rapidly implemented, yet have sufficient detail to support a comprehensive review of the change. Remember that, as described above, these procedures cannot be fully employed if changes are not identified effectively.

In applying the advice and ideas from this book, consider that there is a life cycle to implementing engineering change controls. It includes identifying when to start controlling change in the design process, who approves the controlling systems and how, as well as what steps to take if this does not occur as intended. Decisions must be made on linking the change control system with the company's validation master plan and on how to control costs. Virtually every pharmaceutical company will have to examine how the change control system can be integrated into each stage of plant development: tender, design, construction, commissioning and validation. There must be a constant emphasis on quality assurance. This can be ensured by keeping supporting documentation of a plant that is "as designed, approved and as built," and by submitting vigorously to appropriate audits. The chapters in this book can help you address each part of the life cycle of engineering change control.

Chapter 1 explains that changes are inexorably unavoidable. Historically and economically, we are bound by forces outside the area of our control, with only the knowledge that our business will most certainly be affected by them. Chapters 2 and 3 cover the regulatory response to addressing the change control issue in the U.S. It is important for managers to respond to such features proactively, since these aspects may determine whether or not a company will be allowed, by the appropriate authorities, to continue marketing operations. By necessity, these two chapters have different structures, but they are sure to make useful tools for managers, whether their products are marketed in European or North American jurisdictions. Chapter 4 describes the operational requirements and responsibilities to ensure that change controls are effectively applied and recorded. This chapter covers what you need to do to control change. Chapter 5 provides guidance on how to ensure and demonstrate that your change control system is adequate and working as intended. Those familiar with computers know of their potential weaknesses and the importance of validation when using them in pharmaceutical plants. Chapter 6 addresses the impact of changes on the specific and ubiquitous role of computers in production of pharmaceuticals today and ways of dealing with those changes. Chapter 8 covers how the change control system might be applied on specific projects. This chapter is not necessarily pharmaceuticals-specific, but has been included as an additional part of this book's toolbox to support the overall theme.

Chapter 7 contains useful case histories and anecdotes to illustrate key points, and could provide the reader with a basis for change control training. This should not be regarded as a fully representative compilation of cases: One could keep adding to these case histories as part of the process of improving an organization's corporate memory. To do this, readers will need to continuously keep abreast of new incidents, operational difficulties and accidents caused by changes.

Some may be new to the concept of controlling change or may wish to sample ideas and advice as required to improve existing systems. The chapters are divided to give guidance on managing aspects of change control, which could be brought on stream holistically within a company or in separate stages as required. I hope that the ideas, case studies, and advice will assist in providing the most practical way for you. If you wish to describe any further useful pharmaceutical change control case histories for inclusion in future editions of this book, please forward them to the editor at: simon_turner@fwuk.fwc.com.

The information in this book is believed to be entirely reliable. However the editor, authors and publisher specifically disclaim that any compliance with the approaches and recommendations in this book will make any operations or manufacturing facilities safe, healthy or in compliance with any rule, law or regulation.

References

1. Gillett, J.E., *Hazard Study and Risk Assessment in the Pharmaceutical Industry*, Interpharm, Buffalo Grove, IL, 1996.
2. Institution of Chemical Engineers, Modifications: the management of change, training package 025, IChemE, 1994.
3. Sanders, R.E. *Management of Change in Chemical Plants*, Butterworth-Heineman, Oxford, England, 1993.

Acknowledgments

The editor thanks the friends and colleagues who have helped in writing and contributing ideas and information in this second edition of this book. The first edition of this book would not have been possible without the appropriate opportunity generated by Peter Thomas. Thanks to Validation In Partnership Ltd. for supplying additional useful case histories specifically for this book and Mick Long for preparing some of the accompanying illustrations. I express particular thanks to Steven Vranch for writing the foreword at such short notice. Terry Gilfillan, Malcolm Harrison, Richard Larkin, Bob Adamson, Jon Shattock and Paul Elliott also provided support and encouragement on behalf of Foster Wheeler Energy Ltd. The editor also wishes to thank Deirdre Cranefield and Lynn Harvey of the Reading Foster Wheeler Library for their valuable support in obtaining references used for this work.

The editor thanks Foster Wheeler Energy Ltd. for giving permission and support to publish this book.

The Editor

Simon G. Turner, B.Sc., C.Eng., FIChem.E., is a chemical engineer with more than 18 years experience in the petrochemicals, chemicals, pharmaceuticals, offshore and nuclear industries. He has specialist knowledge and experience in hazard identification, evaluation and analysis, risk management and assessment, and reliability technology.

He holds a degree in chemical engineering from Aston University, Birmingham, U.K. He is a fellow of the Institution of Chemical Engineers (IChemE), is a chartered engineer and retains registration as a safety and loss prevention specialist with the IChemE. He is also a member of the Hazards Forum.

Turner is currently IChemE Subject Groups Forum chairman and Thames Valley Centre chairman. He was an elected member of IChemE Council, serving a 3-year term (1998–2001). He chaired the IChemE Safety and Loss Prevention Subject Group from 1997 until 2000 and was also winner of the IChemE's 1998 Hanson Medal. He was awarded his first patent in October 2003.

Contributors

Mark Hayes, Ph.D.
Genzyme Corporation
Framingham, MA

Ian Holloway, M.Sc.
Medicines & Healthcare products
   Regulatory Agency
   (formerly MCA)
Hertfordshire, U.K.

**Tony J. Margetts, Ph.D., B.Sc.,
F.I.Chem.E.**
Consultant
Wilmslow, U.K.

Leslie Rhubin, M.Sc., RAC.
Genzyme Corporation
Framingham, MA

Kieran Sides
Validation in Partnership, Ltd.
Cheshire, U.K.

Peter W. Thomas, B.Sc.
Genzyme Corporation
Suffolk, U.K.

**Simon G. Turner, B.Sc., C.Eng.,
F.I.Chem.E.**
Foster Wheeler Energy Limited
Berks, U.K.

**Stephen P. Vranch, F.R.Eng., M.Sc.,
C.Eng., F.I.Chem.E.**
Jacobs Engineering U.K. Ltd.
Surrey, U.K.

Table of Contents

1

Change Is Inevitable — Why?

Simon G. Turner

CONTENTS

I. Objective

The English writer Gilbert Keith Chesterton (1874–1936) wrote in his work *Orthodoxy*, "If you leave a thing alone, you leave it to a torrent of change." This may sound paradoxical to those newer to industry, but it is very true indeed. The objective of this chapter is to illustrate this point by giving examples of which facets and nuances of business life are prone to change and to confirm that there is little we can actually do to prevent change. We can, however, take the necessary steps to minimize negative impacts of change on our businesses, provided we can gain an understanding of the more important aspects. How change can be managed effectively and practically is covered in other chapters in this book.

II. The Economics

A flux of economic influence naturally induces changes in competitive organizations. All successful enterprises want to improve profits. This aim will always demand enhancement of a company's plant and systems. These enhancements could occur at any time during a system's life cycle. If the improvement in profits is significant, it is more likely the enhancing change will have accompanying pressures for it to be implemented quickly.

Effective competition is important to business success. Competent company boards will ensure that there are constant pressures to stay competitive or lead the field in key areas of company businesses. They will direct strategic and tactical responses to outside influences, thus, a general force of evolution in industrial practice occurs in virtually every company. Good management is all about making changes to deal with evolutionary or sudden shifts in circumstances. Companies that are traditionally thought of as having good managers tend to be the ones that predict which changes are most likely to occur, and proactively respond before the changes are forced upon them.

External influences that affect the economic viability of a company include:

1. New diseases can become apparent or conditions may reintroduce or favor the spread of a particular disease or condition. Should one of these threaten to become an epidemic, there will be huge pressures to produce new or improved drug treatments (in the case of smallpox, influenza A, AIDS, Ebola, or fish Pfiesteria) or massively increase production of medicines (in the case of influenza C or skin cancer). Life-forms seem to have an amazing capacity for survival. This is apparent in strains of bacteria and viruses, which are known to regularly undergo mutation, becoming resistant to harsher environments and, hence, existing treatments. Therefore, we need to

improve antibiotics to keep ahead of drug-resistant disease. Efficacy of treatments in general application should be regarded as having a half-life; and new treatments can be expected to supersede them rapidly. A state of natural chaos ensures the evolutionary axiom of "survival of the fittest," which applies to organizations as well as organisms.

2. Newly patented treatments and patents running out are annual occurrences for some companies. The impact of losing exclusive rights to a particular treatment could have a major influence on the business. The switch to competing with a flood of generic treatments on the market might mean operational changes also must be dealt with. There is then a need to reduce costs in order to stay competitive. New treatments bring a threat to businesses. Though they could potentially make a lot of money, previous new wonder products, within or closely allied to the pharmaceutical industry, have led to huge problems for some companies in the past. Changing to new, potentially market-leading products could adversely affect business. Consider these examples: Thalidomide,[10,11] Southern Corn Leaf Blight,[10] The Great Cranberry Scare,[10] and The Cyclamate Affair,[10] to name a few. Consult the abstracts of these changes to using new products in Chapter 7 or directly consult the two references; both are very interesting and well written. Another type of problem for new drugs lies in the possibility that a drug may fail to be approved by the regulatory authorities (e.g., U.S. Food and Drug Administration [FDA]) at the latter stage of clinical trials. This is a business setback, which will result in changes to any existing business plans that are in place for the drug that has failed its clinical trial.

3. Management requirements that result from the turnover of staff in positions of influence — or other organizational changes — may affect the economic health of a business.

4. Natural changes or cycles (weather, drought, insect populations, plant diseases) can all affect the demands on a company as well as the supply routes it relies on for key ingredients. Such changes, even dealing with a crisis, could affect you. Corporate responses may involve minor or major changes in the way the company operates.

Management of change always seems to involve the same kinds of problems and also appears to involve the same kinds of approaches to solving them. The key difference across the globe seems to be when different companies recognize that change is necessary or that change is happening with management stimulation. Differences in company culture are also evident when it is recognized that some further changes within the company are required to deal with a monitored change. These further changes might typically be with the intention of:

- Cutting costs and attempting to revitalize dying parts of a business.
- Improving operations (see Chapter 4).
- Ensuring economic success.
- Improving management in the future.
- Dealing with the influence of the international stock market or potential for takeover.
- Ensuring business risk and hazard risk reduction.
- Dealing with potentially fickle customer demands.

Economic influences are probably the strongest forces for change. As employees of a company, having to respond to board directives, we must recognize that resistance to change is, at best, going to slow the organization down and will probably be viewed by senior managers as counterproductive to improving business. Good managers respond positively by being prepared for these changes with suitable systems of working.

III. The Regulatory Demands with Particular Regard to Healthcare Manufacturing Regulations

Chapters 2 and 3 describe the impact of change on regulatory affairs (RA) within organizations. Regulatory authorities do make changes of their own to ensure the validity of particular treatments. These are likely to change the way a company manufactures its products. These changes need to be subjected to controls laid down by the relevant regulatory authorities, allowing a license to market in a particular country. Good manufacturing practice (GMP) advancements also have impacts, as described below.

IV. Changes in Good Manufacturing Practice

A. Concept of current Good Manufacturing Practice (cGMP)

The GMPs adopted by the industry comprise legal and guidance documentation. The regulatory framework in the U.S., the Code of Federal Regulations, is controlled by the FDA. In Europe, the law is addressed by European Community (EC) directives. A similar legal structure exists in other developed countries. The law is supplemented by regulatory group guidance documents, although these are not legal documents. Industry groups (e.g., PhARMA [Pharmaceutical Research & Manufacturers of America] in the U.S.) and professional groups (e.g., International Society for Pharmaceutical Engineering [ISPE] and Institution of Chemical Engineers [IChemE]) also offer GMP guidance. Published papers and inspections by relevant bodies serve as a basis for regulatory development of GMP levels. There is thus a seemingly swirling mass of changing GMP documentation, which effectively represents a current status of GMP (cGMP).

Most businesses have to comply with GMP to obtain a product and manufacturing licence. In the past, companies have been forced into unwelcome takeovers, or even liquidation, by failing GMP inspections (e.g., Fisons in the 1990s).

Over the past 20 years, there has been a steady ramping up of the quality levels applied by the industry in order to avoid GMP noncompliance. In particular, this has increased the level of documentation (validation) that has to be prepared and also the quality of the manufacturing environment. (heating, ventilation, and air conditioning and building finishes, etc.). More recently, there has been a reassessment of the baseline requirements so that a sensible "minimum" level of GMP is established.

This continuing evolution of GMP levels affects the quality of the manufacturing facilities and the amount of staff and documentation required to support them.

B. Demonstrating Change Control for GMP

Change control is also a key element of GMP, since the regulatory authorities require companies to demonstrate that they are in control of changes. Regulatory authorities expect companies to evaluate changes to the manufacturing facility, equipment, process, automation and more for impact on product quality. Chapters 2 and 3 cover this topic in more depth.

For examples of the impact of loss of change control on GMP, consider changing to a different method of cleaning (see Chapter 7, Section XII) and then consider procedures (see Chapter 7, Section XIII).

V. Changes in Safety Practices

There is little doubt that over the years spanning the twentieth century, there have been the most dreadful industrial accidents of history. It is not that companies of our age, within industry, do not care about the hazards associated with their operations, but in order to survive the economic circumstances described earlier, these companies must continue to take calculated risks, appropriate to the demands made by the people in the marketplace. The risks, however, should not be due to changes that have not been subjected to hazard study (see Chapter 7, Section XXVI). While changes themselves can create safety problems, we need to recognize that the ways in which we deal with hazards and carry out our safety management are also changing. Below are just a few examples of such changes.

A. Changes in Hazard Study

New drugs and delivery systems will always be waiting just around the corner for us to discover and develop into new and improved products for better, more cost-effective treatments. It is the very nature of newness that leaves companies just short of the necessary experience to ensure risks are minimized from the outset of a new treatment. Safety specialists have hazard study techniques to identify and manage as many hazards as they reasonably can for a new product, its process, and plant. The techniques that are most applicable to the pharmaceutical industry are well described by Gillett.[1] However, even if we employ all of these techniques, it is usually impossible to identify all hazards that come with manufacturing a product — particularly a new one — no matter how similar the new product is to the old one. There will always be some — potentially painful — experience to be learned from every new aspect of processing to consider and to incorporate into our ways of operating in future.

The sword of technological advancement has a double edge. We may have a new drug projected to be as great of a money-spinner as Zantac™ or Viagra™, but we may have to sacrifice run-of-the-mill processing technology and employ novelty to achieve our aim. Novelty is the riskier side of technological advancement, not only in commercial terms, but especially in terms of safety.

Should processing technology become fundamentally different from what we are used to dealing with, we may also have to develop new hazard study techniques to deal with particular technologies. More recent examples of such changes, or developments, in hazard study techniques are GENHAZ, for identifying hazards associated with releasing genetically modified organisms (GMOs) into the environment, and CHAZOP, for identifying hazards associated with the use of computers in manufacturing operations. Less recent changes in improving plant design philosophy include considering designs that employ the now well-established concept of inherent safety.[4] Kletz[2] describes in detail ways of introducing inherent safety by design. Effective techniques and ideas like these will always make inroads into our ways of doing things, because good safety managers want to (and must) keep up with key developments.

B. Changes in Data and Design Codes

There also will be changes in safety data. Some material safety data sheets (MSDSs) have data change from year to year. For example EH40, an attachment to the U.K. COSHH (Control of Substances Hazardous to Health) Regulations,[3] is revised annually. This means that occupational limits do not have absolute fixed values over time. Codes and standards have changed and will always change with increasing industrial experience. Therefore, the potential for Chinese copying of plant designs, over periods of time, within Europe and the U.S. at least, is difficult to do with complete confidence in safety.

C. Changes in Hazard Analysis

Calculation methodologies are continuously being refined or completely revised. One example of such a change is that blast effects of flammable vapour cloud ignition used to be estimated by comparison with a TNT mass equivalent. This was because in the past, there was a wealth of data available about the blast effects of TNT from war and terrorist attacks, for example. Research within the last 10 to 15 years has confirmed that TNT equivalent blast effects cannot effectively be relied upon where vapour clouds are concerned because ignited flammable vapours behave according to the degree of confinement the vapour is dispersed in. Newer models, such as Baker-Strelow,[5] have been developed and employed widely in industry. Even this has now been superseded.

D. Changes in Containment Practice

The general trend is toward the use of more potent compounds. This creates problems in the working environment, where there is a potential for operators to be exposed to the more active agents. Therefore, there is a need for enhanced containment, including the increased use of isolation technology.

VI. Changes during Design

Changes can occur during the design process. Some of these always seem to creep in after hazard study has been completed. There is, therefore, another, if somewhat subtle, opportunity for changes to be engineered into a plant in an unchecked manner. Companies should not see this type of change as something that should be prevented. It is far more sensible to make such changes to the design before construction commences. Repeating HAZOPs on a later revision of a piping and instrument diagram seems very wasteful, but some companies do it. However, if post-HAZOP design changes are seen and addressed in a systematic manner, there is no reason why drawing developments cannot continue after hazard studies have been completed, more sensibly. The methods for addressing this kind of change vary. Chapter 5 suggests a model system for controlling change and Chapter 8 on project change control considers managing design changes specifically. The core change control system is probably a system suitable for many companies to use for addressing post–hazard study changes. Some organizations may prefer something more streamlined or properly dovetailed into their existing procedures. For example, at Foster Wheeler, design changes are typically managed using special data sheets, detailing the change to each discipline engineer. For changes that do not necessarily impact plant design, there are still procedures put in place to protect the interests of individual projects and the relevant contracted parties. Accordingly, disciplines can easily flag their concerns for project managers to make decisions on what to do about them.

VII. Changes in Software

Software changes may be one of the cheapest business operations to maintain compliance with validation requirements. For example, an operator or line manager may wish to reprogram a programmable logic controller (PLC), which regulates the function of a piece of packaged equipment, to respond to process conditions more quickly. Such changes are relatively easy to make physically. Should a fault be introduced, it is likely to lie in wait (like a time bomb) until certain conditions are met. Some software changes do not wait to cause problems (see Chapter 7). This is one of the reasons why software controls are just as important to control as hardware or process changes. Chapter 6 considers the causes of changes in computer systems and good automated manufacturing practice (GAMP)[12] more fully and provides appropriate advice in dealing with such changes.

VIII. National or Governmental Demands of Industry in General

A. Morality, Codes of Ethics, and Government Policy

The animal lobby has been of concern, in varying degrees, across the pharmaceutical industry. An example for enforced change resulting from animal lobbying was introduction of the law in the U.K. effective from November 1997, which prohibits the testing of any cosmetic on animals. Many cosmetic companies had been using animal testing to support the health and safety of their product on the market. Now, all companies selling cosmetics in the U.K., must comply, which means an alternative way of demonstrating product health and safety had to be found. This meant an upheaval for some, possibly. No doubt the already animal-friendly companies took this change without significant immediate impact. However, animal testing is still necessary in pharmaceutical product development. External pressures may induce changes as to what and how testing is done by pharmaceutical companies, possibly sooner than we think.

Cultural attitudes are very different from country to country and from generation to generation. The Victorian English were great exploiters of colonial territories, bringing access to key resources and immense wealth to the home of the then-great empire. Public opinion and attitudes seem unlikely to support such business practice now. This is an example of a longer-term change, but this kind of change can occur more quickly. Cultural attitudes toward the practice of animal testing differ. Marketing operations may have to change significantly if a company wishes to make an economically successful break into a foreign market; simply as a response to a cultural difference, which can also change with time. Changes to the use of genetic engineering and genetic research in pharmaceutical operations also exist. As research progresses, new questions are raised, by pressure groups or bodies, which will affect an organization's ability to practice certain experiments within a given country. Such pressures can, but do not always, result in changes to political attitudes and, potentially, the law.

B. National Crises

A nation's policies, laws, and economies are not the only influences that can affect business. National unrest or other emergencies may arise. These situations normally occur relatively quickly, making the ability to respond to resulting changes more difficult. After all, a country that has just been subjected to a military coup or earthquake could have been a major supplier of a particular feedstock or ingredient. Not all of these national crises are unforeseeable or sudden. For example, prior to the Gulf War in the early 1990s, the situation in Iraq had been festering for a considerable period, allowing a response to sudden changes to be more manageable. It is not within the scope of this book to give guidance on the national or international crisis issue, mainly because each crisis has to be dealt with on its own merits and circumstances at board level.

C. Public Pressure

Industrial companies are under pressure to lessen the impacts of their operations on the environment. There is no point in griping about the environmental concerns of the public or how they are represented in the media. Every operator has to respond to them. Companies that do not respond well may become the victims of bad press, however justified or unjustified. For example consider the Opren scandal.[11] A nonsteroid anti-inflammatory drug (NSAID) was given the name Opren. It worked better than aspirin and did not have the aspirin side effect of bleeding in the stomach. It was used to enable people with arthritis to live with the disease and avoid surgery. Opren's side effects appeared to include isolated premature deaths in the elderly and sensitivity to sunburn in a small proportion of cases. Media scare reports blamed Opren, and it was withdrawn. The risk of dying as a result of being treated with Opren was about 1 in 25,000; whereas, the risk of dying from the relevant surgery was 1 in 60.

No company can be totally intransigent to public perception of its work. Companies must accept the importance of lessening negative environmental impacts. Without responding to public demands and following up with adequate representation in the media, the whole business operation may be threatened beyond what is reasonably comprehensible for a given public concern. These days, companies have to make very rapid changes to their operations — or even implement changes to their management systems — because of pressures from the public. An example of this is where effluent discharges might have recently increased by a factor of two or three, but were still well below the allowable limits. Poor publicity on this issue would more than likely mean that a company would be forced to make a reduction in effluent discharges. Such changes may or may not actually have very much impact on the outside world. Indeed companies may see little point in making changes from a technical standpoint; but the changes need to be made more often than not.

IX. Company Standards, Locally and Internationally

Improving corporate success is an essential qualification for economic survival. In striving to overcome side effects, poor efficacy, and costs of treatment, advances in medical practice are constantly being made, most of them published. New technology introduces obsolescence into current practices. Obsolescence is a feature of design and operating practice that companies need to weed out of their standards and procedures. My conversations with retired engineers suggest that even when the pace of development was relatively slow, companies were not always effective in dealing with obsolescence (see Chapter 7, Section XXVI).

A. Results of Research and Development

The very nature of research means that new and unknown territories have to be explored. Laboratories might be considered a special area, possibly exempt from change control procedures because of the small quantities of experimental materials involved, but this must not be regarded as the case. A prime example is the use of fume cupboards, where changes are constantly being made. New, more potent compounds, some made by potentially runaway reactions, may be under scrutiny within a fume cupboard. In addition, there will always be changes to solvents used in fume cupboards, which may mean that the flammable atmosphere ignition protection on the electrical equipment is inadequate; at least some of the time. Research and development changes are not confined to the laboratory. Pilot plants are frequently built for establishing the impacts of scaling up new processes. Once the pilot plant is built, the very nature of pilot plants dictates that changes will be made to investigate which parameters can be used to best improve its performance. The changes might be in materials, conditions, or equipment. Similarly, many manufacturing operations now employ multiproduct or general-purpose facilities, which also need to be reconfigured to make new or different products. In all cases one must exercise strict change controls and ensure that comparisons with the original hazard studies done on the pilot plant are used to initiate new hazard studies when necessary.

Technological advances must be the most unavoidable and inevitable change in industry. Without the introduction of newness to products the most powerful driving force behind making sales is lost. There are few people who will make a point of buying last year's treatments, when it is known that hospitals and consumers are actively looking for improvements in efficacy or value. Technology gives us at least one of these improvements and sometimes both.

X. Maintenance

Sanders[6] amply covers how changes occur as a result of maintenance in the general arena of chemical manufacturing. Examples of some of the factors that influence the degree to which maintenance introduces change are:

- Poor communication or procedures
- Inadvertent change to the design (see Chapter 7, Section IV)
- Variation in replacement parts or components
- Slow, incipient changes that occur during each maintenance operation (see Chapter 7, Section VIII)

In order to compete, each company will have to address the economic realities of responding to technological developments and market needs. This will certainly include, at some stage, examination of its internal methods of production and design standards to satisfy new demands. Most companies appear to adopt a single standard, worldwide, for the design of their facilities. These appear to be based on the most restrictive practices in the countries where they operate. Standardization might make dealing with changes easier for them.

XI. Date-Related Functions

Most readers will recall the period when some computer applications were likely to be affected by the transition to the year 2000. This phenomenon arose because shorthand, two-digit numbers were employed to represent the actual four-digit year. Some software was unable to handle the date being represented as zero-zero in the year 2000. It was one of those changes you could see coming. There is no escaping this potential for certain key date changes to create havoc for our business systems, even now. It is also doubly important to realize that some safety and critical business systems are likely to be affected by this date-related function on an ongoing basis. If your systems of work are date reliant (which in pharmaceuticals manufacturing is virtually certain), you have probably witnessed minimal (if not aggressive) management response to the "millennium bug" within your company. Remember that the rollover to the year 2000 was not the only problem date. There are problems associated with software, with date-related functions, the inability to handle leap years correctly, 366 days per year, 29 days in a month, and so forth.

Regrettably, the date-related function problem is not one that can be addressed by the issue of a corporate policy or statement. There was a saying that the year 2000 problem was like peeling an onion; the more you peel the more you reveal, the more you reveal, the more you cry. The date-related functions problem is general to all industries, so is not within the scope of specific technical advice covered within this book. However, there are many published works that can assist with going beyond the change to the year 2000.[7-9] Even if you already have addressed date-related functions issues internally, it may be beneficial to consult the key advice and recommendations of these and other changes to the year 2000 works, to indicate how date-related functions can be more effectively managed.

I hope this chapter has shown that a torrent of change is continually impacting business operations. The extent of impact is not the same for each source of change. As long as we recognize that changes will occur — and that we must manage them effectively — our organization's likelihood of survival and its enhanced performance will improve.

References

1. Gillett, J.E., *Hazard Study in the Pharmaceutical Industry,* Interpharm, Buffalo Grove, IL, 1996.
2. Kletz, T.A., *Plant Design for Safety — A User Friendly Approach,* Hemisphere, New York, 1991.
3. HMSO, *Control of Substances Hazardous to Health Regulations,* Her Majesty's Stationery Office, 1994.
4. Kletz, T.A., What you don't have can't leak, *Chemistry & Industry,* SCI, London, 287, May 6, 1978.
5. CCPS, *Guidance on Techniques for Explosion Modeling,* AIChE, 1996.
6. Sanders, R.E., *Management of Change in Chemical Plants,* Butterworth-Heineman, Oxford, England, 1993.
7. Hugo, I., *Managing Year 2000 Conformity: A Code of Practice for Small and Medium Enterprises,* DISC-PD2000-2, British Standards Institution, London, 1997.
8. Henderson, J. and Davidson, G.I., Safety and the year 2000, *Health and Safety Executive,* Her Majesty's Stationery Office, 1998.
9. BSI, *A Definition of Year 2000 Conformity Requirements,* DISC-PD2000-1, British Standards Institution, London, 1997.

10. Lawless, E.W., *Technology and Social Shock*, Rutgers University Press, New Brunswick, NJ, 1977.
11. Elmsley, J., *The Consumers' Good Chemical Guide*, Corgi, U.K., 1994.
12. GAMP, *Good Automated Manufacturing Practice*, International Society of Pharmaceutical Engineers, 2003, www.ispe.org/gamp.

2

Regulation of Change Control for Drugs and Biologics — U.S.

Mark Hayes and Leslie Rhubin

CONTENTS

I. Introduction

This chapter has been revised extensively due to numerous changes in regulations and approaches to change control implemented by the U.S. Food and Drug Administration (FDA) in recent years. Over the past decade, the U.S. has enacted legislation aimed at reforming the way the FDA reviews and approves products with an eye toward streamlining the approval process. The "GMP Initiative" announced by the FDA in 2002 will undoubtedly continue that evolution. The FDA has engaged in revising change reporting regulations and guidelines primarily to avoid discouraging innovation by manufacturers who, for example, may desire to improve processes, but decline to do so because of onerous change reporting requirements. With reference to postapproval change reporting, Section 116 of the 1997 Food and Drug Modernization Act (FDAMA),[11] the Code of Federal Regulations (CFR), and a variety of guidance documents (all detailed later in this chapter) provide requirements and guidance for making and reporting manufacturing changes to an approved application and for distributing the products made with these changes. Despite the volume of information supplied and examples enumerated in these documents, the specific situations manufacturers encounter often fail to neatly classify into the change levels (e.g., minor, major) and reporting mechanisms (e.g., prior approval, CBE-30, annual report) outlined in them. Typically, an organization's regulatory personnel will be asked to interpret the regulations and guidance and, if appropriate, contact the agency to ultimately determine the most appropriate reporting mechanism for a given manufacturing change, or set of changes.

We will initially discuss how we approach change control internally in the context of a global organization, and the critical need for global harmonization in this area. The remaining sections of this chapter focus on U.S. legislation, regulations and guidance documents, issued over the past decade, that address evaluating and reporting manufacturing changes for marketed products to the FDA.

A. Globalization Requires Regulatory Affairs Integration in the Change Control System

Pharmaceutical manufacturers typically operate in global markets and often manufacture multiple products at multiple facilities. This creates challenges to regulatory affairs personnel in providing assessments of change reporting requirements in the various markets, making the associated filings to the authorities, and notifying customers. A solely reactive role will not suffice in establishing an effective change control system. Ensuring worldwide regulatory compliance relies on actively integrating the regulatory function with operational objectives. Given the realities of business globalization and the slow progress toward global regulatory harmonization, a change control system needs coordination among personnel responsible to multiple regulatory authorities, and efficiency in that coordinated effort, to be truly effective. Figure 2.1 illustrates one process design for assessing worldwide

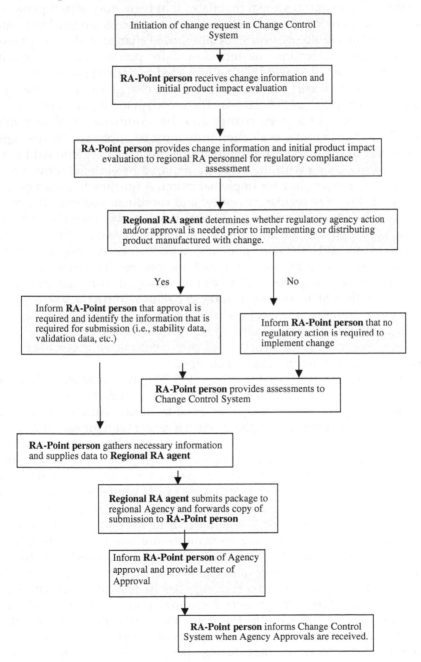

FIGURE 2.1
Worldwide regulatory assessment process.

regulatory requirements for anticipated manufacturing changes and providing the requirements to the corporate change control management system. In this design, the regulatory evaluation process is initiated at the time that the change control system mandates that regulatory affairs personnel evaluate the manufacturing change request. The process can also be initiated to evaluate the regulatory impact of a proposed change, and the regulatory burden can be factored into the decision-making process before committing resources to moving forward with a proposed manufacturing change.

The regulatory affairs function can improve change control efficiency at more than one point in the process. The most apparent points include the initial assessment of a given change and the worldwide regulatory filing plan. If the change under evaluation truly must be implemented, the aggregate impact of the market-specific reporting mechanisms identified by regulatory affairs, along with the evidence required by each authority, can be factored into the timeline for implementation. A timeline that incorporates the expected time of regulatory review and coordinates submissions to the various jurisdictions to minimize the potential time between the first jurisdiction approved and the last jurisdiction approved reduces the burden on the operations and inventory planning departments. As long as effective tools are used to communicate and track the progress of the change implementation plan, unexpected delays can be managed. For changes that are desirable, rather than necessary, initial regulatory impact assessments are critical to evaluating the benefits of implementing the change. In some instances, the benefits that a company expects to realize from a given change can be all but nullified if the filing plan is mismanaged, regardless of the quality of evidence supporting the change.

There is an additional risk incurred when filing plans are not fully developed and monitored in concert with the change implementation strategy. During the approval cycle, companies need to ensure that they have inventory made in compliance with the product's registration for each market. At times, this means manufacturing a product that incorporates the process change for markets in which approval was granted, while holding inventory made using the previous process for specified markets. When inventory stocks are depleted, manufacturers may chose to manufacture more material using the previous process, or withhold the medicines from patients in those markets yet to be approved — the latter choice having readily apparent negative implications. Choosing to manufacture one product using more than one process strains the manufacturer's quality unit and adds to company expenses, as the potential for mix-up during manufacture, testing, release, and distribution increases, along with the costs to manufacture multiple "versions" of the same product. Factoring the manufacturing and inventory changeover strategy into the regulatory filing plan, and regularly communicating approval progress and inventory supply, can reduce the instances of inventory shortage during change implementation, as options to potentially expedite approvals and reallocate inventory can be evaluated quickly.

The following example serves to illustrate the concepts outlined above. A manufacturer of several marketed protein therapeutic products derived from recombinant DNA (rDNA) technology wants to begin production of a new therapeutic protein using the same facilities and personnel that manufacture its current licensed products. In this instance, many regulatory authorities want assurance that the introduction of the new product will not result in cross-contamination or mix-up of the products. Table 2.1 provides the regulatory reporting mechanisms that the manufacturer must follow to maintain regulatory compliance in the jurisdictions that license its currently marketed products. In each jurisdiction, the necessary waiting period following submission or agency approval must be obtained before distributing licensed product lots made after introducing the new product into the manufacturing area. From a cursory review of the regulatory assessments, one might conclude that it will take three to four months to gain all the approvals necessary. However, further analysis reveals that, at best, it may take 10 months to obtain approval in the last jurisdiction where submission takes place, provided the timing of the various submissions is well managed. In this example, the Israeli submission would not be filed until the FDA approval letter is received — up to 6 months after the U.S. filing. Note that in Europe, no submission is required before distributing lots of the licensed products after introducing the new product in the manufacturing area. As this example

TABLE 2.1

Example of Worldwide Regulatory Obligations for Manufacturing a "New" Product in a Licensed Multiproduct Facility

Jurisdiction (Agency)	U.S. (FDA)	European Union (EMEA)	Canada (Health Canada)	Israel (Ministry of Health)
Reporting Mechanism	Changes being effected — 30 days	No reporting required	Notifiable change	Prior approval submission
Agency Review Period (as defined in regulations)	30 days	None	90 days	Not mandated
Comments	Product may be shipped "at risk" pending formal FDA approval, not fewer than 30 days after the receipt of the supplement by the FDA; formal FDA approval may take 6 months	This is a GMP issue subject to evaluation at on-site inspection	Information and material must be filed prior to the institution of the change; unless a written objection is received from the branch within 90 days, the manufacturer may proceed with the change	In this example, submitted information includes a copy of a formal FDA approval letter; approval timeline may be 3–4 months

illustrates, the most expeditious regulatory filing plan (the order and timing of submission to each agency) emerges only after the individual assessments are evaluated against one another. In the ideal scenario, regulatory affairs personnel use the assessment information to construct a timeline for the filings, which minimizes the amount of time required to gain the necessary approvals worldwide. The timeline is communicated to the quality assurance and operations teams, who then determine how much of each product to make, or have on hand, before introducing the new product into the manufacturing area. If all goes according to plan, the introduction can be seamless.

In general, product manufacturing and development arrangements are not always straightforward, as alluded to previously. Whatever the legal relationship (joint venture, development partnership, bulk API supplier, contract, shared, divided, or short supply manufacturing arrangement), the product licensee is responsible for maintaining regulatory compliance in the markets that regulate the pharmaceutical. In the global arena, the licensed party may be mandated by the attendant jurisdiction's regulations. Under these circumstances, where the parties involved in the manufacturing and development relationships have different region-specific roles, the change control system must be maximized for cross-functional communication and integration. As such, when entering relationships in which multiple parties bear some responsibility for manufacturing, development and licensing activities, the parties involved would be prudent to clearly define the change control communication system prior to signing the contract or technical agreement.

B. Other Considerations

Regulatory compliance applies to all stages of the pharmaceutical life cycle, during premarket and development, as well as postapproval. In the U.S., with regard to products in the clinical development and premarket stages, the mechanisms and reporting requirements are not as clearly articulated as those for postapproval changes. Nevertheless, sponsors of products in the development stage need to evaluate proposed chemistry, manufacturing and controls (CMC) changes for their potential to adversely impact the safety and efficacy profile established through completed animal and clinical studies. Proactive communication with the product's reviewers at the FDA, regarding CMC changes under consideration during development, is well advised. At the time of market application filing, the FDA may refuse to file an incomplete application due to insufficient description of the manufacturing process, including:

- Incomplete data demonstrating equivalency to clinical trial product when significant changes in manufacturing processes or facilities have occurred.

- Failure to describe changes in the manufacturing process, from material used in clinical trial to commercial production lots.[20]

In the U.S., when a market application exists, or is pending, while active clinical study of a product continues under the Investigational New Drug Application (IND) — which allows an unapproved drug to ship in interstate commerce for clinical studies — both applications must be maintained. The maintenance procedures are independent of one another. Therefore, proposed changes must be filed separately following application-appropriate pathways, provided the changes are intended, or affect the CMC conditions specified in each application. Once again, globalization adds another layer to this scenario in that each region will have its own set of regulations to apply when clinical study and pending market approval occur simultaneously.

C. Relief in the Form of Harmonization

On the whole, harmonization efforts undertaken by regulators, in conjunction with industry, will provide the basis for reducing complexity in maintaining global regulatory compliance. Through consensus building, the multitude of jurisdiction-specific requirements may be reduced to those that provide the best means for assuring that postmarketing changes do not adversely impact a product's quality, safety, or effectiveness. Already, considerable effort has been expended by the International Conference on Harmonization of Technical Requirements for Registration of Pharmaceuticals for Human Use (ICH) to discuss and develop consensus on the technical requirements for product registration, as well as provide a common document format for that registration. Initially formed to focus on registration requirements for Europe, Japan, and the U.S., the ICH steering committee (composed of members associated with the regional regulatory authorities and industry organizations) recognized the need to expand its communication and dissemination of information with non-ICH parties.[21] Also, at its fifth international conference in 2000, the ICH steering committee "identified post marketing activities as a future area where increased regulatory cooperation can help to contribute to the enhancement of the protection of the health of citizens on a more international basis."[21]

Even if global harmonization of regulatory requirements and expectations lags behind the emergence of the global pharmaceutical marketplace, the effort that each agency expends in clearly defining expectations and mechanisms for maintaining regulatory compliance will benefit the pharmaceutical manufacturer, as well as provide a basis for future harmonization efforts. As mentioned at the beginning of this chapter, the U.S. has enacted legislation over the past decade that reforms the way the FDA reviews and approves products. Ultimately, the driving force behind the U.S. legislation is in concordance with the ICH's goal of ensuring that safe and effective medicines are developed and registered in an efficient manner, in the interests of the consumer and public health.

II. Risks of Noncompliance: Auditing and Documentation

There are only two windows that the FDA and other regulatory authorities have into confirming controls associated with the products they regulate: CMC sections of marketing applications (and INDs) and on-site inspections. Adequate descriptions of processes and controls are essential for approval of any marketing application for a pharmaceutical product. If companies want flexibility in changes anticipated in their CMC, the relevant CMC sections should be explicit in identifying areas where such flexibility is assumed. The challenge to any change control system will come during inspections. If changes are made, they should be captured in a formal documentation system. We use engineering change requests (ECRs) and document change requests (DCRs) for documenting changes to processes and standard operating procedures (SOPs), respectively. Appropriate review and sign-off by relevant departments assures that changes are actively managed, validated when necessary, and reported to regulatory authorities when appropriate. Signatories on these change control documents assume responsibility for assessing the impact of any change. Change control databases and revision histories should be subject to audit on a regular basis.

III. Current Regulations and Guidance for Evaluating and Reporting Manufacturing Changes to the FDA

For drugs and biologics approved through new drug applications (NDAs) or biologics license applications (BLAs), the mechanisms for reporting manufacturing changes in the U.S. are stipulated in FDA regulations pertaining to change reporting: 21 CFR 314.70 (drugs)[22] and 601.12 (biologics).[23] There are essentially three levels of reporting: prior approval supplement (PAS), changes being effected (CBE), and annual report, based on the potential for adverse impact on the product in question.

1. Changes that have a substantial potential for adverse impact on the product must be approved through a PAS. The timeline for approval of a PAS mandated by the Prescription Drug User Fee Act (PDUFA) is 4 months.

2. Changes with a moderate potential for impact on the product can be reported through a CBE supplement. These changes can be implemented 30 days following receipt of the supplement (CBE-30) or

immediately (CBE-0). However, formal approval letters are not typically sent to the sponsor until well after the specified time period elapses.

3. Changes with minimal potential for impact can be reported in the annual report for each product and can therefore be implemented immediately.

The FDA has issued numerous guidance documents for assessment of change reporting. All of these guidelines are posted on their Web site (www.fda.gov). In spite of explicit descriptions for how to report particular changes or types of changes offered by the FDA through scale-up and post-approval changes (SUPAC) and other guidance documents, there are many circumstances where reporting mechanisms are unclear. In such cases, sponsors are advised to contact the FDA for specific guidance.

A. History and Revisions to Change Reporting Statutes and Regulations

The FDAMA,[11] signed into law on November 21, 1997, amended, among other statutes, those pertaining to the requirements and procedures for implementing and reporting manufacturing changes to approved NDAs, abbreviated new drug applications (ANDAs), and BLAs. In response to the FDAMA mandate defined in Section 116 of the Act, the FDA issued a proposed rule in 1999 to update and replace the regulations pertaining to reporting of manufacturing changes.

These regulations are defined in 21 CFR 314.70 (for drugs) and 21 CFR 601.12 (for biologics). The 314.70 regulations were initially drafted as part of the NDA rewrite in 1985. This rewrite created the now familiar three-tiered system for reporting changes: supplements requiring prior approval, supplements not requiring prior approval, and annual reports. Then, as now, determination of how to report using these mechanisms was based on the significance of the change and its potential to affect the safety or effectiveness of the product. Following the implementation of 314.70, in an attempt to define consistent guidelines for manufacturers, the Center for Drugs Evaluation and Research (CDER) convened the SUPAC task force, which oversaw the issuance of guidance documents for categorizing changes for reporting manufacturing changes for drugs.

B. SUPAC: CDER Guidance Documents for Specific Drug Product Dosage Forms

The SUPAC guidelines were issued based on the dosage form subject to the change. This approach was based on the assumption (and FDA and industry experience) that different dosage forms were at different risks for the impact of changes to their manufacture, composition and control.

1. SUPAC-IR: Immediate Release Dosage Forms

The first of the SUPAC guidelines, published in November 1995, was for immediate release dosage forms (SUPAC-IR).[1] SUPAC-IR provided, for the first time, detailed guidance not only on how to report changes to IR products, but the documentation expected to support specific changes by category. The major categories covered in this guidance were as follows:

- Changes in components and composition (excipients)
- Manufacturing site changes
- Changes in batch size (scale-up or scale-down)
- Changes in manufacturing (equipment or process)

This and subsequent SUPAC guidance documents also define levels of change, based on their likelihood to adversely impact the quality or performance of the product. Level 1 changes are those that are considered unlikely to have any detectable impact. Level 2 changes are those that could have a significant impact. Level 3 changes are those that are likely to have a significant impact on quality or performance. For each level, SUPAC-IR recommends specific testing to be performed to evaluate the change, based on variables such as therapeutic range, solubility and permeability (as an indication of expected extent of absorption). In addition, the extent of stability experience is taken into account in the assessment of stability data requirements to support these changes. The requirements are diminished if the sponsor has a "significant body of information" with respect to stability data. A "significant body of information" is defined as "likely to exist after 5 years of commercial experience for new molecular entities, or three years of commercial experience for new dosage forms." However, regardless of the defined requirements, the inherent stability characteristics are also likely to be critical in decisions about how to assess changes.

Changes in components and composition are defined quite specifically in SUPAC-IR with respect to excipients by function. Cases are laid out based on permeability and solubility, and assist in defining the need for *in vivo* bioequivalence documentation. The latter may be waived in the presence of an acceptable *in vivo/in vitro* correlation.

Site changes in SUPAC-IR are evaluated based on whether the new site of the operation is within a single facility, within a contiguous campus, or at a different campus. A contiguous campus is a "continuous or unbroken site or set of buildings in adjacent city blocks." This somewhat ambiguous definition is considered more accurately with respect to whether the same equipment, SOPs, environmental conditions, and personnel will be used. Clearly, the concern with respect to potential impact on product quality is increased when any of these circumstances change. In addition, an obvious expectation for any new site is the completion of a satisfactory current Good Manufacturing Practice (cGMP) inspection by the FDA. For overseas manufacturing sites, the FDA has not yet accepted a mutual recognition procedure with other inspectorates; therefore, an inspection by the FDA is still required even if other authorities have completed inspections of the same site.

Changes in batch size (scale-up or scale-down) are evaluated using, among other factors, the 10X model (i.e., scale changes greater than 10X are considered to be more significant). The other factors include details on the equipment and SOPs used, and assumes cGMP compliance.

Manufacturing changes are considered with reference to changes in equipment or the actual process. Frequently, changes are not made in isolation. When changes are combined under any of these categories, it is advisable to communicate with the agency to assess appropriate reporting criteria and supporting documentation required for review.

In 1997, the FDA responded to a growing number of questions from manufacturers about how to specifically interpret the SUPAC-IR guidance by publishing a questions and answers document.[2] At that time, the FDA also reconsidered guidance on two issues affected by the original SUPAC-IR. Specifically, changes in stand-alone packaging operations and stand alone analytical testing sites were both permitted to be reported using a CBE supplement, provided satisfactory cGMP compliance could be assured through a recent inspection. Additional questions addressed details in each area of change defined in the original guidance. Manufacturers should carefully review this question and answer document when evaluating changes for which circumstances are not explicitly defined in the SUPAC-IR guidance.

2. SUPAC-MR: Modified Release Dosage Forms

The FDA/CDER issued the second SUPAC guidance document, applicable to modified release dosage forms, in October 1997 (SUPAC-MR).[3] This guideline essentially follows the same format as SUPAC-IR, in that three levels of change are defined within the following categories:

- Changes in components and composition (excipients)
- Manufacturing site changes
- Changes in batch size (scale-up or scale-down)
- Changes in manufacturing equipment
- Changes in manufacturing process

This guidance differs from SUPAC-IR in the category of changes in components and composition in distinguishing release-controlling excipients from non-release-controlling excipients. In addition, distinction is made between extended-release and modified-release drug products. Changes in manufacturing equipment and process are presented in separate sections of the guidance. As with SUPAC-IR, however, the FDA cautions against extrapolating from changes defined in this guidance and recommends discussion with the agency for evaluation of multiple changes implemented at the same time.

Other specifics differentiate SUPAC-MR from SUPAC-IR. SUPAC-MR provides details recommending additional dissolution documentation for extended- and delayed-release products that are either not compendially defined or described in the application. In assessing changes in components and composition, the manufacturer must justify the declaration of a specific excipient as release or nonrelease controlling. In the evaluation of equipment changes, SUPAC-MR defines a requirement for validation if the new equipment is not identical, in every respect, to the original manufacturing equipment used in the approved application. SUPAC-MR also introduces the concept of a "critical equipment variable," "critical manufacturing variable," and "critical processing variable" (defined by whether the equipment or manufacturing change is likely to be critical to drug release).

3. SUPAC-SS: Nonsterile Semisolid Dosage Forms

The FDA published a third manufacturing change guidance in May 1997 for nonsterile semisolid dosage forms (SUPAC-SS).[4] Again, this guidance categorizes changes on the basis of components and composition, site changes, batch size, and equipment and process. The scope of this guidance includes creams, gels, lotions, and ointments intended for topical use. SUPAC-SS differs slightly from the SUPAC-IR and SUPAC-MR guidelines in specifically addressing multiple changes in addition to single changes. Issues associated with these types of dosage forms, (i.e., manufacturing of compounds that exist in two or more phases and the impact of combining phases and sequence of addition of the active pharmaceutical ingredients) are central to the philosophy taken in SUPAC-SS. In addition, special considerations for this class of products are discussed with respect to the value and limitations of *"in vitro* surrogate tests"* in assessing the impact of changes. SUPAC-SS provides recommendations on design and performance of *in vitro* release tests and statistical analyses to be used in comparison testing. The CDER's position on *in vitro* release testing was articulated in SUPAC-SS as follows:

1. *In vitro* release testing is a useful test to assess product "sameness" under certain scale-up and postapproval changes for semisolid products.

2. The development and validation of an *in vitro* release test are not required for approval of an NDA, ANDA, or AADA, nor is the *in vitro* release test required as a routine batch-to-batch quality control test.

3. *In vitro* release testing, alone, is not a surrogate test for *in vivo* bioavailability or bioequivalence.

4. The *in vitro* release rate should not be used for comparing different formulations across manufacturers.

Following the issuance of SUPAC-IR, SUPAC-MR and SUPAC-SS, the FDA published guidance documents to assist industry in evaluating specific manufacturing equipment changes for these products. These manufacturing equipment addenda[5,6] were designed to be used in conjunction with the existing SUPAC guidance documents. These addenda define, by category of operation, specific types of equipment by class and subclass. However, in the SUPAC-IR/MR guidance, the FDA again cautioned against over-interpretation of the addendum, particularly with reference to "critical equipment variable" equipment used in the manufacture of modified release products. Equipment described in the SUPAC-IR/MR addendum is listed by function or unit operation, using the following categories:

- Particle size reduction and separation: milling, cutting and screening
- Blending and mixing: diffusion, convection and pneumatic
- Granulation: dry or wet, high or low shear, extrusion, rotary, fluid bed or spray drying
- Drying: direct heating or indirect conduction or radiant; static, moving, fluidized or dilute
- Solids beds; lyophilization; gas stripping
- Unit dose operations: abletting, encapsulating or powder filling
- Soft gelatin capsule operations
- Coating, printing and drilling

The accompanying Manufacturing Equipment Addendum Draft Guidance for SUPAC-SS was issued by the FDA in December 1998. The categories of equipment and unit operations defined in the SUPAC-SS addendum are as follows:

- Particle size reduction and separation; impact, attrition, compression and cutting
- Mixing: convection, roller and static
- Emulsification: low or high shear
- Deaeration
- Transfer: passive or active
- Packaging: holding, transfer, filling and sealing

C. BACPAC: Bulk Active Chemical (Pharmaceutical Ingredient) Postapproval Changes

The SUPAC initiative was designed to address changes to commonly developed drug product dosage forms. More recently, the CDER and the Center for Veterinary Medicine (CVM) have begun to expand this concept to manufacturing changes for drug substances manufactured by chemical synthesis. Only the first of these guidelines has been published: BACPAC I, Intermediates in Drug Substance Synthesis.[7] This guidance covers changes associated with manufacture of bulk actives up to the point of the "final intermediate," defined as "the last compound synthesized before the reaction that produces the drug substance." The referenced reaction assumes formation of a covalent bond (e.g., not generation of a salt form of the compound). BACPAC I is limited in scope. It does not address synthetic peptides, oligonucleotides, radiopharmaceuticals, or compounds that are not subject to chemical synthesis, but are derived from "natural sources" or "procedures involving biotechnology." Changes associated with biologics and biotechnology products are addressed in separate guidelines (see Section entitled "Manufacturing Changes for Biologics").

The types of change addressed in BACPAC I are site, scale and equipment changes, specification changes, and manufacturing process changes. BACPAC I provides a decision tree for evaluating whether a change falls within the scope of the guidance. The requisite conditions are whether equivalence is evaluated at the stage of the intermediate and where in the process the impurity profile is analyzed. Two underlying assumptions support limiting the scope of BACPAC I. The first is that "the risk of adverse change is generally acknowledged to be greater when a modification occurs near the end of a drug substance manufacturing process rather than at the beginning." Note that this is most certainly not the case for biologically derived drug substances (see below). The second underlying assumption is that the drug substance itself is well-characterized (not in the same sense originally used to define "well-characterized" biologics prior to the subsequent and more appropriate application of the term "specified" to these biological products). BACPAC I also excludes changes in source material for natural source intermediates used in "semi-synthetic" processes.

In evaluating changes to bulk active intermediates, BACPAC I assumes the critical attributes to be the structural analysis as assessed by physical parameters, and the relative impurity profile. This guidance does not provide stability testing recommendations. However, as noted in BACPAC I, there may be circumstances in which stability analysis should be performed, either on the drug substance or on drug products derived from same.

BACPAC I goes to some lengths to explain the concept of "equivalence of impurity profiles." A manufacturer is in a better position to assess the impact of manufacturing changes if impurities are monitored at critical stages in the process (i.e., testing of "isolated intermediates"). It is important to recognize the limitations of available analytical procedures in performing impurity testing for equivalence. The FDA recommends testing at least three consecutive postchange batches and comparing these data to historical results for the same intermediates or drug substance (consecutive batches are recommended for this as well). The guidance also cautions against comparing data to older retained samples where impurity levels may have increased over time. For human drug substance intermediates, equivalence in impurity profiles is defined in BACPAC I as follows:

- No new impurity in the intermediate is observed at >1.0% or above the "qualification threshold" as defined in ICH Q3A.[8]
- Each existing impurity (including residual solvents) is at its stated limit or is at or below the upper statistical limit of historical data.
- Total impurities are within the stated limit or are at or below the upper statistical limit of historical data.
- New residual solvents are at or below the levels recommended in ICH Q3C.[9]

D. PAC-ATLS: Postapproval Changes — Analytical Testing Laboratory Sites

The FDA/CDER has issued a guidance document to define circumstances under which reduced reporting (i.e., from a prior approval supplement to a CBE supplement) may be acceptable for changes in analytical testing sites (PAC-ATLS).[10] Specifically, a change in analytical testing site may be submitted using a CBE mechanism if all of the following apply:

- The test method(s) approved in the application or methods that have been implemented under 21 CFR 314.70(d) are used.
- All postapproval commitments made by the applicant relating to the test method(s) have been fulfilled (e.g., providing methods validation samples).
- The new testing facility has the capability to perform the intended testing.
- The new testing facility has had a satisfactory current good manufacturing practice (cGMP) inspection within the past 2 years.

Absent any of these conditions, implementation of a new analytical testing site must be reported through a prior approval supplement.

E. FDAMA and Manufacturing Changes

As noted previously, the FDAMA[11] amended the statutes pertaining to the requirements and procedures for implementing and reporting manufacturing changes to approved NDAs, ANDAs, and BLAs in 1997. Section 116 of FDAMA amended the Federal Food, Drug, and Cosmetic Act (FD&C Act) by adding section 506A (21 U.S.C. 356a) to the statutes. On June 28, 1999, the FDA published in the *Federal Register* a proposed rule revising 21 CFR 314.70 (64 FR 34608) and proposing changes to 21 CFR 601.12 to harmonize the drugs and biologics regulations on this topic.[12] In addition, the FDA made available at that time a draft companion guidance entitled Changes to an Approved NDA or ANDA (64 FR 34660).[13]

The essential purpose of the FDAMA provision was to consolidate progress that had already been made (e.g., SUPAC, the 601.12 rewrite, and similar efforts) with respect to defining and, where appropriate, reducing reporting requirements for manufacturing changes. The summary language in FDAMA Section 116 was as follows:

> With respect to a drug for which there is in effect an approved application under section 505 or 512 or a license under section 351 of the Public Health Service Act, a change from the manufacturing process approved pursuant to such application or license may be made, and the drug as made with the change may be distributed, if: (1) the holder of the approved application or license (referred to in this section as a "holder") has validated the effects of the change in accordance with subsection (b); and (2)(A) in the case of a major manufacturing change, the holder has complied with the requirements of subsection (c); or (B) in the case of a change that is not a major manufacturing change, the holder complies with the applicable requirements of subsection (d).

The FDAMA Section 116 defines a major change as "a manufacturing change that is determined by the secretary to have substantial potential to adversely affect the identity, strength, quality, purity, or potency of the drug as they may relate to the safety or effectiveness of a drug." A major change is more specifically defined as a change "in the qualitative or quantitative formulation of the drug involved or in the specifications in the approved application or license" or a change that requires "completion of an appropriate clinical study demonstrating equivalence of the drug to the drug as manufactured without the change." However, the statute also provides for any other "type of change determined by the secretary by regulation or guidance to have a substantial potential to adversely affect the safety or effectiveness of the drug." Obviously, each manufacturer, in coordination with the FDA, must make their own judgements about what may fall into the category of "substantial potential" for impact. Section 116 proceeds to describe regulatory alternatives for other changes not deemed to be major and directs the FDA (the secretary) to define categories for change reporting that are now described in the current 21 CFR 314.70 and 21 CFR 601.12 and associated guidelines. The FDA was careful to clarify the meaning of the statutory phrase "validating the effects of the change" in the proposed rule issued in 1999.[12] Specifically, the FDA does not necessarily take "validation" to have the same meaning as the corresponding term defined in the cGMP regulations provided in 21 CFR 210 and 211. This allows the FDA and manufacturers more flexibility to review cGMP validation documentation on site as part of the inspection process, where appropriate, rather than as part of a submission.

Ultimately, the final rule that issues in response to FDAMA Section 116 may supercede prior guidelines such as SUPAC if inconsistencies exist with regard to reporting requirements. However, as of the beginning of 2003, the final rule had not been issued. The 1999 proposed rule offered the following revisions to the existing regulations for manufacturing changes:

- Addition of definitions in 21 CFR 314.3 and 600.3 for the terms "specification" and "validate the effects of the change"
- Reformatting of 21 CFR 314.70 and 21 CFR 601.12
- Substantial addition of descriptions of changes and reporting requirements in 21 CFR 314.70

F.   Expedited Review of Manufacturing Changes

The FDA has recognized that there may be circumstances in which an expedited review of manufacturing supplements is appropriate "if a delay in making the change described in it would impose an extraordinary hardship on the applicant."[19] Extraordinary hardship is defined as the following:

- Public health need: Events that affect the availability of a drug for which there is no alternative
- Catastrophic events such as explosion, fire, or storm damage
- Events that could not have been reasonably foreseen, and for which the applicant could not plan.
- Agency need: Matters regarding the government's drug purchase program, or federal or state legal or regulatory actions, including mandated formulation changes or labeling changes

G.   Manufacturing Changes for Biologics

Changes to biologics licensed under the Public Health Service Act (PHS Act) are regulated under 21 CFR 601.12. However, some biological products, for historical reasons, were approved under NDAs and, therefore, subject to the change regulations under 21 CFR 314.70. In 1997, the FDA issued guidance documents for reporting biologics changes. Two guidelines were issued. The first, issued by the Center for Biologics Evaluation and Research (CBER) was for licensed biological products, including whole blood, blood components, source plasma, and source leukocytes.[14] A second guidance[15] was issued concurrently, by both CBER and CDER, for "specified biotechnology and specified synthetic biological products." What necessarily preceded the latter guidance was the elimination of the establishment license application (ELA) for specified products.[16] In 2001, CBER issued a change reporting guidance specifically for human blood and blood components intended for transfusion or for further manufacture that replaced the 1997 guidance specifically for whole blood, blood components, source plasma, and source leukocytes.[17]

However, the 1997 biologics guidance remains in place for other biologics that are not specified or subject to the 2001 guidance.

The biologics guidance documents reference 21 CFR 601.12 and, for specified biologics, also reference 21 CFR 314.70(g) for biotechnology products regulated under the FD&C Act. As for drugs, three levels of change are assumed, on the basis of whether they have a substantial, moderate, or minimal potential to have "an adverse effect on the identity, strength, quality, purity, or potency of the product as they may relate to the safety or effectiveness of the product." For the purposes of these guidance documents, changes with substantial potential to adversely affect the product are considered major changes with reference to the FDAMA definition above.

Major changes (with substantial potential for adverse effect) require prior approval supplements and are defined with identical examples in both 1997 biologics guidance documents. These include process changes in both cell culture and purification operations, such as extension of cell culture growth time and changes in columns or buffers used in purification. Of particular note and often misunderstood, reprocessing is subject to approval through a prior approval supplement in the absence of a previously approved reprocessing protocol. Also subject to prior approval are any process or analytical method changes that affect specifications. Scale-up requires prior approval and is not subject to the 10X rule applied to drugs. Any change in scale is potentially significant, although minor changes in fermentation batch size may be reportable by CBE-30 if specifications are not affected. Also unlike SUPAC, any change in excipients or other variables in the dosage form are subject to prior approval.

Other changes requiring prior approval common to both 1997 biologics guidelines include changes in acceptance criteria for reference standards and expiry extensions not supported through real-time data from an approved stability protocol (accelerated data is typically of limited utility for biologics). Manufacturing site changes, conversion of single-product to multiproduct facilities, and even changes in location of specific operations within a facility typically require prior approval. Multiproduct issues are critical for biologics, and addition of new products or changes in locations of operations are subject to prior approval, primarily if contamination or cross-contamination are of potential concern. The proven existence of validated cleaning and changeover procedures can allow for reduced reporting in these situations.

A few facility-related changes are uniquely identified as prior approval changes for biologics that are not specifically noted in the biotechnology guidance as such. These include significant water for injection system changes and major construction; heating, ventilation, and air conditioning; or operation location changes that impact environmental control.

A limited number of changes are identified as examples in both 1997 biologics guidelines that are subject to CBE-30 reporting. These include addition of duplicated process chains (i.e., additional bioreactors or columns) or changes in the number of pieces of equipment with no change in process parameters. In addition, a change in analytical testing sites and introduction of additional products into already-approved multiproduct facilities can be reported through CBE-30 supplements. Again, a few changes are uniquely identified as CBE-30 changes for biologics that are not specifically noted in the biotechnology guidance. These include automation or addition of computer controls over process steps (with no change in process), changes in vial fill volume, changes in responsible individuals (likely a holdover from the older concept of the responsible head), and certain facility modifications, including to water systems. Finally, addition of release tests/specifications and tightening of intermediate specifications are reportable through CBE-30.

A unique regulatory option that has been available to biologics manufacturers since 1996 is the use of a comparability protocol.[18] Comparability protocols, which define predetermined acceptance criteria and additional testing requirements to assess and support specific changes, are submitted as prior approval supplements. Upon approval, changes that are assessed through such protocols can then be submitted and approved through a reduced reporting mechanism (i.e., CBE-30 or annual report). This mechanism is typically only appropriate for products and manufacturing processes for which there is substantial experience.

A series of changes common to both 1997 biologics guidelines are identified as annual reportable. These include minor changes to analytical methods; certain harvest or pooling procedures; replacement of reference standards using SOPs defined in the license; tightening of specifications and addition of alternative test methods for reference standards; replacement of working cell banks; changes in storage conditions for intermediates; changes in shipping conditions; addition of tests, time points, or tightening of specifications in stability protocols; trend analyses of release data; and changes in simple floor plan of manufacturing facilities. Many of these are annual reportable only if established protocols exist in the license, so it is critical to define such protocols prior to submission of a license application for any new product. Additional facility and equipment-related changes are specifically added as annual reportable for nonspecified biologics. The requisite conditions for such changes remaining annual reportable are that there are no changes in environmental quality or process parameters.

IV. Moving Forward in Managing Change

What can we expect with regard to further guidance on reporting manufacturing changes? As noted under FDAMA and Manufacturing Changes, the final rule for regulation changes proposed in response to FDAMA has not yet been issued. In fact, it is not clear whether a final rule will be issued, since there are ongoing discussions on change reporting within the FDA and among industry stakeholders (e.g., 2002 GMP Initiative). In what will undoubtedly be of help to biologics manufacturers, ICH has taken up the topic of comparability in an attempt to harmonize regional expectations.

However, even with more and better guidance documents, there will always be room for interpretation. Manufacturers must take responsibility for knowing their products and processes, and recognize not only when to implement and report changes, but how best to assess their impact. The importance of open communication with regulatory authorities cannot be understated when addressing these complex issues.

References

1. FDA, *SUPAC-IR: Immediate Release Solid Oral Dosage Forms Scale Up and Post-Approval Changes: Chemistry, Manufacturing and Controls, in vitro Dissolution Testing, and in vivo Bioequivalence Documentation,* Center for Drug Evaluation and Research, Rockville, MD, 1995.
2. FDA, *SUPAC-IR: Questions and Answers about SUPAC-IR Guidance,* Center for Drug Evaluation and Research, Rockville, MD, 1997.
3. FDA, *SUPAC-MR: Modified Release Solid Oral Dosage Forms Scale Up and Post-Approval Changes: Chemistry, Manufacturing and Controls, in vitro Dissolution Testing, and in vivo Bioequivalence Documentation,* Center for Drug Evaluation and Research, Rockville, MD, 1997.
4. FDA, *SUPAC-SS: Nonsterile Semisolid Dosage Forms Scale Up and Post-Approval Changes: Chemistry, Manufacturing and Controls, in vitro Release Testing and in vivo Bioequivalence Documentation,* Center for Drug Evaluation and Research, Rockville, MD, 1997.
5. FDA, *SUPAC-IR/MR: Immediate Release and Modified Release Solid Oral Dosage Forms Manufacturing Equipment Addendum,* Center for Drug Evaluation and Research Rockville, MD, 1999.
6. FDA, *SUPAC-SS: Nonsterile Semisolid Dosage Forms Manufacturing Equipment Addendum (Draft Guidance),* Center for Drug Evaluation and Research, Rockville, MD, 1998.
7. FDA, *BACPAC I: Intermediates in Drug Substance Synthesis; Bulk Actives Postapproval Changes: Chemistry, Manufacturing, and Controls Documentation,* Center for Drug Evaluation and Research, Rockville, MD, 2001.
8. ICH, *Q3A: Impurities in New Drug Substances,* ICH, Geneva, Switzerland, 1996.
9. ICH, *Q3C: Impurities: Residual Solvents,* ICH, Geneva, Switzerland, 1997.
10. FDA, *PAC-ATLS: Post-Approval Changes — Analytical Testing Laboratory Sites,* Center for Drug Evaluation and Research, Rockville, MD, 1998.
11. Food and Drug Administration Modernization Act of 1997 (Pub. L. 105-115).
12. Federal Register, *Supplements and Other Changes to an Approved Application,* Vol. 64, No. 123, pp. 34608-34625, June 28, 1999.
13. FDA, *Guidance for Industry: Changes to an Approved NDA or ANDA, Questions and Answers,* Center for Drug Evaluation and Research, Rockville, MD, 2001.
14. FDA, *Changes to an Approved Application: Biological Products,* Center for Biologics Evaluation and Research, Rockville, MD, 1997.
15. FDA, *Changes to an Approved Application for Specified Biotechnology and Specified Synthetic Biological Products,* Center for Biologics Evaluation and Research, Rockville, MD, 1997.

16. *Federal Register, Elimination of Establishment License Application for Specified Biotechnology and Specified Synthetic Biological Products*, Vol. 61, No. 94, pp. 24227-24233, May 14, 1996.
17. FDA, *Changes to an Approved Application: Biological Products: Human Blood and Blood Components Intended for Transfusion or for Further Manufacture*, Center for Biologics Evaluation and Research, Rockville, MD, 2001.
18. FDA, *FDA Guidance Concerning Demonstration of Comparability of Human Biological Products, Including Therapeutic Biotechnology-derived Products*, Center for Biologics Evaluation and Research, Rockville, MD, 1996.
19. FDA, *MAPP 5310.3: Requests for Expedited Review of NDA Chemistry Supplements*, Center for Drug Evaluation and Research, Rockville, MD, 1999.
20. FDA, *SOPP 8404.2: Refusal to File Guidance for Biologics License Applications*, Center for Biologics Evaluation and Research, Rockville, MD, 2002.
21. ICH, *The Future of ICH — Revised 2000*, ICH Steering Committee, 9-11 Geneva, Switzerland, 2000.
22. Code of Federal Regulations, Title 21, Part 314, *Applications for FDA Approval to Market a New Drug*, U.S. Government Printing Offices, Washington, DC.
23. Code of Federal Regulations, Title 21, Subchapter F, Biologics, Part 601, Licensing, U.S. Government Printing Offices, Washington, DC.

3

A Regulatory View — EU

Ian Holloway

CONTENTS

I. Introduction

The European Union (EU) is a dynamic expanding array of nationalities, cultures and histories, and progress has been made toward pan-European legislation and control for medicinal products. Expansion of the EU will continue for the foreseeable future, and differences in interpretation and application of legislation will undoubtedly continue between member states. This chapter was written from the author's perspective as a good manufacturing practice (GMP) inspector in the U.K. It covers inspection and validation issues and touches upon marketing authorization (MA) issues. This edition contains additional information on recent controls relating to Transmissible Spongiform Encephalopathy (TSE) risks and additional controls relating to purchasing and formulation.

II. Brief History and Recent Developments

For many years, the U.K. Medicines Control Agency (MCA) has inspected overseas manufacturers who supply the U.K. market. This work will continue after the ongoing merger with the Medical Devices Agency to become the Medicines and Healthcare Products Regulatory Agency (MHRA). A similar approach to overseas sites has been taken by the U.S. Food and Drug Administration and other inspectorates such as the Australian Therapeutic Goods Administration (TGA), resulting in some duplication of GMP inspections and the need for manufacturers to satisfy different inspectorates. Progress has been made on reciprocal recognition of inspections between the EU and several countries such as Australia, but hopes of such agreements between the EU and U.S. still appear to be a distant dream.

The continuing expansion of the EU will introduce additional pharmaceutical manufacturers seeking to export to the richer member states. Several EU inspectorates are providing advice to these countries on establishing GMP inspection systems and regulatory frameworks.

Many new applications are being received, naming third-world countries such as India. Although many of these applications are for generic, nonsterile products, there are some applications for sterile products. Reciprocal agreements with these countries are not presently on the horizon. These sites are likely to demand an increasing share of inspection and regulatory resources, and the demands of working in such countries should not be underestimated.

III. The Change to Mutual Recognition Agreements

One of the key changes over the past few years has been the development of mutual recognition agreements (MRAs) between the EU and other countries. Some EU states have had bilateral agreements with other countries for several years (e.g., Canada and the U.K.). These agreements have been replaced by new MRAs, covering all EU states. The activities of the Pharmaceutical Inspection Cooperation Scheme (PICS) have facilitated this.

The detailed content and scope of MRAs vary from agreement to agreement and each should be carefully studied. Some agreements differ in coverage of human and veterinary products and have had different implementation dates. For products not covered by the legislation of both parties, inspections can still be requested. New systems are normally needed for reporting defective products.

IV. How the Change to MRAs Could Affect You

In the U.K., and in many other EU states, MRAs have caused a reduction in overseas inspection activities, resulting in some reduction in costs and product launch times to manufacturers.

Probably the most significant MRA would have been between the EU and U.S. Unfortunately, the transitional period ended, and no decision was taken to extend the process, which has been and still is effectively stalled. The two-way alert system between the parties remains in operation.

V. Future Changes — The Increasing Scope of GMP

The European Commission (EC) is progressing regulations that will ensure that all starting materials used in medicinal products, unless specifically exempted, are made in accordance with GMP. This is likely to include all active ingredients and some excipients and controls on investigational medicinal products (IMPs) used in clinical trials.

Controls on clinical trials are well advanced, and this will be the first area to be enacted. Key legislation includes directives 75/319/EC,[1] 81/851/EC,[2] now codified as directive 2001/83,[3] and the clinical trials directive 2001/20/EC.[4] The latest implementation date for the new controls is May 2004, by which time all IMPs will have to be manufactured in accordance with GMP, and sites will be subject to GMP inspections. There already have been some advisory inspections of these activities by the MHRA. The framework for formal inspections of clinical trials is currently being established throughout the EC.

VI. Notification Procedures, Forms and Documents

A. MAs

The EU system for variations to marketing authorizations has been operational since June 1995. In the U.K., this has been applied to medicinal products authorized through mutual recognition and centralized procedures together with U.K. national procedures. Some other states have solely applied this system to mutual recognition procedures. Notable exceptions have included homeopathic products and clinical trials certificates.

B. Types of MA Variation

Variations are defined as Type 1 or Type 2. Type 1 variations are minor, and have to be processed within 30 days. Type 2 variations are complex, and have to be processed within 90 days. Engineering changes could affect either type.

An example of a Type 1 variation would be a change in the coating weight of tablets. A list of over 30 categories of Type 1 variations is provided to applicants. If the dissolution profile changed, this would become a Type 2 variation. Thus, the introduction of a new coating machine may suggest that the dissolution profile, from pilot studies, would not cause a change, but production lots may prove otherwise. Type 1 studies require supporting data but do not require an expert report. It is always possible that the regulatory assessor may categorize a variation as Type 2 even though the applicant has classified it as Type 1.

Examples of Type 2 variations include major formulation changes, new containers for sterile products, new sterilization methods, and other complex changes. These variations require an expert report as well as supporting data. In these cases, full validation of equipment and process is needed together with stability studies.

There is not yet one EU-wide application form for these changes. The U.K. uses a mutual recognition document for such applications and a modified version for national applications.

VII. Transmissible Spongiform Encephalopathy (TSE) and Regulatory Controls

TSEs include Bovine Spongiform Encephalopathy (BSE) in cattle and Kuru and Creutzfeldt-Jakob Disease (CJD) in humans. Iatrogenic transmission has been reported in both humans and animals with variant CJD (vCJD), first described in 1996. This has resulted in the application of increasingly stringent regulatory controls in Europe with wide regulatory inspection and change control implications.

Regulatory measures apply to materials of animal origin, which are used for the preparation of active substances, excipients and materials, and of reagents used in production such as culture media, Bovine serum albumin, and enzymes. In addition to those mentioned above, items such as machine lubricants will be considered during GMP inspections.

Several routes are theoretically possible for achieving compliance, although in practice there may be only one route available to manufacturers. For some products, reformulation to use nonanimal sources is available. An example of this would be the use of magnesium stearate from vegetable sources in tablet formulations. If it is not possible to switch to nonanimal materials in the formulation, there are two main routes available to producers. First, for some ingredients, certificates of suitability for commonly used materials are available from the European Directorate for the Quality of Medicines (EDQM). Second, for risk materials for which an EDQM certificate is not available, information on the source and processing must be submitted and assessed by an authority such as the MHRA. This is handled as an application to vary the MA.

Site inspections will normally verify that materials in stock have acceptable certification and that there are no unsupported materials in storage areas, production and laboratories. Purchasing lists will also be compared with production documents to ensure there is consistency for product names and codes. It is by no means uncommon to find differences that could result in the purchase of unapproved materials. Particular care is required in items that are sourced from brokers rather than the original manufacturer.

Contract manufacture is a further risk area, and comprehensive technical agreements will need to be available. It is also important at overseas sites to ensure that TSE evaluation not only covers products for Europe but also covers all materials that could contaminate product-contact items. Risk materials for the home market may represent a contamination threat due to the problems in demonstrating inactivation of potential contaminants.

The change control system should ensure that purchasing functions are fully included and consistent names and codes are used throughout the company. Do raw material codes differentiate between different suppliers of the same ingredient? Material such as gelatin may be delivered in bulk to silos or supplied in reusable bulk containers. It is vital to ensure that these materials are not switched to unapproved sources and then changed back to pharma-grade materials.

VIII. Manufacturer's Licenses

In the U.K., each manufacturing site is inspected and approved before being listed on a manufacturer's license (ML). Such licences are renewed every five years, although a system of continuous licensing is possible. The ML lists the general classification of products that may be manufactured and assembled on the site. Further controls are exerted by listing specific processes and details of potent and biological ingredients. The ML does not list information such as marketing authorization numbers, building names or production line descriptions.

A. Changes Affecting MLs

The types of engineering change that would result in the need for a variation to the ML would be the introduction of a new category of product or a new type of process. Examples might include the production of tablets on a site previously making liquids or beginning to pack a product that was previously processed by a contract packer.

B. How Changes Affect MLs

When variations such as those mentioned above are notified to the MHRA, they are passed to the local GMP inspector for comment. This triggers a GMP inspection for a major change that would take place prior to approval. Routine GMP inspections take place every two years for satisfactory sites, and planned future developments are discussed during those visits. Manufacturers are also encouraged to discuss plans with their local inspector as early as possible. This informal system has worked well and has frequently included the discussion of plans with manufacturers outside the EU.

Further controls on the content of the ML include a check for correct content during routine GMP inspections. The ML is also reviewed by the inspector at the time of renewal. In the U.K., this information is computerized and local inspectorate offices can retrieve this information and print out current licence documents.

C. Site Master Files (SMFs)

The second major document involved in this area is the SMF. This document is produced by the manufacturer for each site and is issued to the inspector prior to the inspection. The SMF contains information such as factory layouts, organization charts, and details of quality systems. Engineering information is included such as details of heating, ventilation, and air conditioning systems and designs of water plants. SMF documents are useful in planning inspections and writing reports. It is usual for these documents to be updated before a routine reinspection and following major changes such as the introduction of a new packing area.

Updating an SMF in response to such changes is usually carried out by reissuing the document and sending a copy to the inspector. The entire contents should be checked to ensure that the document remains current.

For non-EU manufacturers, there is no ML, and a current SMF is an essential tool in preparing for an inspection. Guidelines for the preparation of an SMF are sent to a site prior to the inspection, and the completed document should be returned to the inspector well in advance of the inspection. The SMF is a concise document, often no more than 25 to 30 pages long. It is essential that the official guidelines are followed in preparation of this document; adaptation of an existing document is likely to be unacceptable.

Discussions have taken place on the format of an EU SMF, which would replace the present national documents. The content of this has not yet been decided, but it could be linked to a formal requirement for manufacturers to update the SMF following defined site alterations.

IX. Validation, Revalidation and All That Is Entailed

A. New Equipment

Following a change such as the introduction of a major piece of equipment or new line, the GMP inspector will visit the site and review the entire validation process. This will normally include design qualification (DQ), installation qualification (IQ), operational qualification (OQ) and process qualification (PQ). Production documents and analytical data normally would be reviewed for three production-sized batches. Associated data, such as staff training and cleaning validation, would also be included.

B. Changes to Sterile Areas

For a new or modified sterile products area, there will be a considerable validation effort needed for areas such as air systems, water plants, environmental monitoring, and media fills. The inspectors will examine these areas in detail and may request initial operational data to be forwarded to their offices. Inspectors will also discuss training programs and records for new operations, and ensure that standard operating procedures (SOPs) have been produced.

EU inspectors operating in overseas countries, including the U.S., often find that environmental limits for sterile areas do not comply with the EU GMP guide.[5] Annex 1 of the guide is for sterile products, and a new guide with several changes was published in 2002. Limits for nonviable particles are available for "at rest" and "in operation" states. There are also recommended limits for microbial contamination including active air sampling, settle plates, contact plates, and operators' gloves. Inspectors frequently receive SMFs for new overseas sites and find that the overseas site does not comply with these limits. A copy of the guide should be available on each site that plans to supply medicinal products to the EU.

C. Changes to Process Cleaning and Sterilizing Systems

The increasing use of clean-in-place (CIP), or sterilize-in-place (SIP), and aseptic processing such as blow-fill-seal (BFS) are examples where engineering control and evaluation of steam quality are critical. Inspectors will discuss steam quality checks such as dryness fraction and noncondensable gases. Many other issues are included, such as dead-legs, dual-sourcing of steam, and potential for blocking of orifice plates. The resources needed to validate these areas to acceptable standards are still frequently underestimated.

D. Changes to Computer Systems

The march of the microprocessor into equipment, services, and monitoring systems is creating an increasing need for computer validation systems. Although the year 2000 passed without major disasters in the industry, there is no slowdown in the introduction of new computer systems. Validation and control of computerized systems are being inspected in greater detail by all EU inspectorates. Inspectors will be using documents such as Good Automated Manufacturing Practice (GAMP)[6] to help with the planning and carrying out of computer audits.

Any change to software, however small, should result in a new version number. In some cases, new versions of software have been installed in a pharmaceutical plant, unbeknownst to the user, by a well-meaning service engineer. The use of laptop computers by visitors, or for updating programmable logic controller systems, requires extra care in virus control. It will be necessary for company staff members to supervise visiting engineers during service visits and to have enough knowledge and experience to evaluate the work being done.

The decreasing life of computer hardware creates further problems. A new computer may seem identical, but is the basic input output system (BIOS) version the same? Could a new motherboard with faster memory safely replace an earlier board? Is the system adequately shielded from radio transmissions from neighboring buildings and vehicles? With new systems, it may be worth buying identical spares for key components at the time of purchase — they may become obsolete and unobtainable just a few months later.

The increasing use of the Internet has been accompanied by an increased number of virus attacks on company computer systems. While firewalls may protect obvious risks, the increasing use of personal digital assistants, wireless systems and dial-in services will continue to provide significant challenges. For clinical trials, the use of interactive voice recognition systems

(IVRS) for code control, stock movements, and other uses will present a significant validation challenge. Many of these will present risks to GMP systems, and security is likely to be discussed in more detail during GMP future inspections. These are just a few computer-related issues. For complete guidance on the impacts of changes to computerized systems and how to resolve them, see Chapter 6.

Validation may be contracted out where in-house resources are limited. However, the specification for this work is critical and often subject to careful cost scrutiny. It may be prudent to use independent consultants to advise on the specification for such contracts where in-house knowledge is limited or involved in other areas of the project.

Revalidation is an area where different approaches exist between the U.S. and the EU. For critical equipment such as a steam sterilizer, the EU inspector normally would expect both heat distribution and penetration studies to be carried out annually. Studies should include empty chambers and a range of production loads. The identification of a worst-case load for such studies needs great care. Where long lengths of tubing or complex filter housings are present, adequate air removal may only be demonstrated by the use of biological indicators. The duration and depth of any air removal stages will also need to be carefully considered.

If there are several sterilizers of the same age, capacity, and manufacture on one site, they cannot be considered as identical for initial validation or revalidation purposes. Individual studies will be needed for each machine.

X. Maintaining Compliance and Auditing

The requirements and needs of change control systems are discussed elsewhere in this book. The cornerstone of compliance is an audited, effective, change control system combined with good communication between engineering, regulatory, production, and quality assurance staff.

A. Compliance and the Qualified Person

1. Controlling Compliance

The existence of a regulatory compliance department, seen at most U.S. sites, is less common in most EU pharmaceutical companies. This is probably because legal action against manufacturers is less common in most EU states, in the event of noncompliance. Compliance is based upon the areas mentioned above and is coordinated locally by the qualified person (QP).

The control of compliance within EU states has a significant input from the QP, who is a key member of the quality assurance team and is named on the ML. Directive 75/319/EEC lays down the requirements for human products, and directive 81/851/EEC covers veterinary products. Member states within the EU must ensure that the requirements are met by administrative measures or by codes of practice. In the U.K., a code of practice is employed.

2. QP Responsibilities

QPs have defined key duties, which include ensuring that manufacturing and analytical testing methods have been validated and that change control is an essential part of this process. QPs also need to communicate well with the regulatory affairs department, which is often located at another site. The release of individual batches in accordance with the marketing authorization is a key responsibility of the QP. QPs are also involved in self audits and the release of medicinal products imported into the EU.

Some QP duties can be delegated, but the QP is expected to be conversant with manufacturing conditions. This may mean that QPs have to audit overseas sites and some suppliers. On large manufacturing sites, it is common for several QPs to be employed. EU GMP inspectors expect QPs to actively release batches on a routine basis. If QPs work in isolated areas or move to another department, some retraining may be needed if duties are resumed.

There are disciplinary procedures available if QPs fail to fulfill their duties. In the U.K., such actions would involve the professional bodies who maintain the register of QPs. In serious cases, the licensing authority, who can delete names from the manufacturer's license, is notified.

B. The Audit Process in Relation to the Change Control System

Audits of manufacturers can be considered as corporate audits, customer and supplier audits, self audits, quality system audits or official GMP audits.

1. Corporate Audits

Corporate audits are more frequently found in larger multinational groups and frequently include validation, change control and compliance with MAs. Recently, corporate audit teams have been downsized as part of cost-containment exercises, which is considered a retrograde step. Mixed audit teams, consisting of full-time corporate auditors and factory staff from sister plants, can perform a useful training function. Another major benefit of such audits is the production of company-wide quality standards, which, in some cases, exceed national GMP standards. The production of company-wide procedures for change control and computer validation is usually driven and coordinated by corporate audit teams. The short-term cost benefits of disbanding such teams are frequently followed by fragmentation and an overall reduction in company GMP standards.

2. Customer and Supplier Audits

Customer and supplier audits can provide similar benefits to corporate audits. This is especially true where a major company helps a smaller — or inexperienced — company increase their standards and become an accredited supplier. This may include areas such as help with validation protocols, purchase and loan of equipment, and even supplying staff to work in the other site. This type of help may involve considerable resources when the smaller or weaker site has no experience of GMP, but perhaps holds a patent on a prized process or product.

3. Self Audits

Self audits should also include elements such as change control, validation steps and notification to regulatory affairs. The policy within the MHRA is for inspectors to assure themselves that these audits are regularly carried out, but they normally do not look at any of the findings or content. In some small sites, it is difficult for the person conducting these audits to be independent, because the person may include a large portion of his or her normal work. One solution is to employ a consultant in such cases.

4. Quality System Audits

Quality system audits are becoming more common as pharmaceutical producers achieve ISO 9000 registration, or comply with other quality standards. The most common level is compliance with ISO 9000, with many sites currently upgrading from the 1994 version to the 2000 standard.[7] Originally, the auditors used in these situations had little or no knowledge of pharmaceuticals, but the situation has improved as auditors with industrial experience have been hired. The MHRA inspection section has been accredited to ISO 9000 for several years, and other EU inspectorates are working toward accreditation or are already there. For further information on ISO 9000 definitions, see Chapter 6.

5. GMP Audits

In the U.K., routine GMP audits of manufacturers are carried out every two years. For large sites, the work is usually divided into a series of visits rather than covering all activities during one inspection. If standards on the site are marginally acceptable the site is reinspected at a shorter interval.

Inspection fees are levied by the MHRA, based upon site activities, staff numbers, and whether the site produces human products, veterinary products, or both. Site activities are classified as sterile, nonsterile, or assembly (packing only). Staff numbers are defined as less than 10 (minor), 10 to 59 (standard), 60 to 249 (major), or 250 and above (supersite).

The normal format of inspection can be divided into three phases — an introductory session, a main inspection, and a summary session.

1. An introductory session is used to check ML details and confirm that progress on previous deficiencies has been completed. Future plans are also discussed, including engineering and building projects. A schedule is agreed for the inspection.

2. In a main inspection, the majority of the time is spent visiting the production and laboratory areas. This includes any area that the inspector considers to have GMP content ranging from drains to databases, sterile areas to spectrophotometers. Inspectors see the concentration of activities at one site becoming more widespread with perhaps all tablet production taking place at one site. An increasing amount of time is spent looking at major new production areas and validation packages.

3. Following the inspection, a summary session takes place, at which any deficiencies are reported to the company. If the company considers any points to be incorrect, this is a chance to put the matter right. Deficiencies are classified as "critical," "major," or "other" by the inspector. Critical deficiencies could result in a product recall or formal action against the manufacturer's license or marketing authorization. Both critical and major deficiencies are related to sections in the GMP guide. Findings are formally sent to the site in a letter, and a written response is requested. Following a satisfactory response, a closeout letter is sent to the site confirming general compliance with GMP.

If formal action is needed, the inspection findings are sent to the Inspection Action Group (IAG). This group includes legal staff, medical staff, product-licensing staff, and inspectors. The patient risk, supply situation, and routes for improvement are all evaluated. For a U.K. company, the ML could be suspended, varied, or revoked. For an overseas product, there is no ML for the site, and the normal course of action would be taken against the marketing authorization.

At present, the inspection findings within the U.K. are confidential. Inspectors in the U.K. are required to sign the Official Secrets Act. Inspection reports are not available from the MHRA at this time, but reforms are expected in this area over the next few years. The U.K. government has published a "white paper" relating to freedom of information, and there is an intent to release more data into the public domain. The MHRA issued a consultation letter (MLX 292) in December 2002, with options for changing or repealing Section 118 of the Medicines Act 1968. At present, the disclosure of information relating to the licensing of medicines is a criminal offence. Over the next few years, it should be possible to obtain some inspection findings in the same way that information is available from the FDA.

Inspection deficiencies found during official GMP audits of overseas sites are often related to long lines of communication and the lack of technical agreements. This situation can be worsened if manufacture and packing is carried out on three or more sites that include contractors. Problems frequently found include incorrect formulation or expiry dates and unapproved packing materials or printed text. The increasing use of virtual companies is also a challenge to inspectors, and this seems set to increase. A contract regulatory affairs organization in the U.K. could be working for a U.S. corporate regulatory affairs office in Atlanta for a product made in Chicago, packed in Puerto Rico, and imported to the EC via Switzerland.

References

1. EC Directive 75/319.
2. EC Directive 81/851.
3. EC Directive 2001/83.
4. EC Directive 2001/20.
5. *Rules and Guidance for Pharmaceutical Manufacturers and Distributors,* Her Majesty's Stationery Office, London, 2002.
6. Good Automated Manufacturing Practice (GAMP), *Supplier Guide for Validation of Automated Systems in Pharmaceutical Manufacture,* ISPE, Tampa, Florida, 2002.
7. ISO 9000:2000. U.K. supplier BSI Standards, 389, Chiswick High Road, London, W4 4AL.

4

The Operational Phase

Kieran Sides

CONTENTS

I. Introduction

Changes made during the operational phase of a manufacturing facility are often ill thought out, inconvenient, disruptive, and at best, frustrating. For these reasons alone, they should be avoided unless absolutely essential. Another reason for leaving well alone is the cost. It is universally accepted that this is the most expensive time for making any alterations (Figures 4.1 and 4.2).

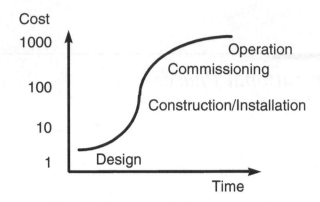

FIGURE 4.1
The relative cost of change.

FIGURE 4.2
"It will cost too much to change now — so it will have to stay as it is."

In many cases, particularly those involving older facilities and operations, changes are brought about by legislative amendments or necessary improvements in working practices. These obviously cannot be helped. However, all too often, needless changes are made to newer plants that could probably have been avoided by putting more effort into deciding what exactly was required in the first place. What may be considered a time-consuming and exhausting process at design stage could pay dividends in the long run, as potential problem areas are envisaged and eradicated before they are built in.

This chapter concentrates on the types, causes, and impacts of change, controlled and otherwise, during the operational phase of a new pharmaceutical manufacturing facility. It could just as easily refer to an oil drilling rig, a car production plant, or an office block because the principles are the same — to ensure that all aspects of the facility operation are well known, well managed and well documented.

II. Control Requires Documentation

It is no use whatsoever having a satisfactorily operating facility if all the knowledge of what keeps the facility in this state is entirely in the heads of those responsible. Human beings are fickle creatures; we have numerous habits that are totally incompatible with routine. We fall ill, we go on holiday, we move to new pastures, we die. So we instigate contingency plans for just such inconveniences.

> For a start, Bill and Eric never have their holidays at the same time; it's not allowed. And Bill's falling ill when Eric is away is tantamount to conspiracy — even sabotage. It's just as well young Henry is a bright lad and can cover in their absence. But Henry, who also knows he's a bright lad and has already been accepted for a new, higher-paid position down the road, is at this very moment on his way to hand in his notice. In a week's time there will be nobody available who knows anything about the job.

If the documentation systems are not in place to enable a replacement, temporary or otherwise, to take over in a situation such as this, then the previously satisfactorily operating facility ceases to be.

We are also blessed with worse habits than those mentioned above. We are creative, we are inventive, we find better ways of doing things. We like to show initiative. We change things, not necessarily because things need to be changed, but because we *can* change them. Even with the best will in the world, all these things might, and probably will, happen. The most we can hope to achieve is to ensure that if changes do occur, they do so in a controlled manner.

What is needed is a process by which the effects and impacts of all these changes are thought out in advance, agreed on by all appropriate parties, and thoroughly documented. The emphasis here should be on the word "documented," because perhaps our worst habit of all is that we forget. Changes, when they go unchecked, will get out of hand; when they also go uncommunicated, they can be downright dangerous.

Without routine, there can be no manufacturing process. Without strict adherence to formal operating procedures and preventative maintenance and calibration programs, there can be no such thing as routine. Furthermore, none of these can be kept up to date without an effective change control system.

III. Preventive Maintenance

Preventive maintenance should be the only policy adopted by those responsible for the upkeep of a facility. The alternative, breakdown maintenance, should not even be considered as an option. Without preventive maintenance, control of plant operations is purely temporary, resource planning cannot be relied on, trends cannot be easily identified, and the stocking of spare parts is, quite frankly, guesswork.

No amount of careful planning will rule out breakdowns altogether, but they will be minimized by programming in the necessary preventive measures. By their very nature, breakdowns cannot be foreseen, and immediate attention may be required. In some instances, it may not be possible to wait for a proposed change to be formally documented and run the round of reviews, approvals, and authorizations. Therefore, the change control system must be flexible enough to allow for such changes to be recorded after the event, albeit at the earliest opportunity following completion of the remedial work.

The action taken may be in the form of a temporary fix just to get the plant back up and running again, and not intended as a permanent remedy. But these temporary fixes have an awful habit of becoming permanent, without the necessary controls. So, whenever a change involves this course of action, the change control documentation should include a date by which the temporary fix is to be removed and replaced by the recommended and agreed permanent alternative.

IV. One System for All

The change control system must be fully operational at facility handover stage to ensure that all changes following the completion of the construction project are captured. This means that all those responsible for the day-to-day running of the facility, namely, production, quality, maintenance, and technical support personnel, are fully conversant with the system. They must be formally trained, and training records must be kept to reflect this. Also, the training should not be just a one-off session; effective change control is the very essence of good facility management. All personnel should be required to attend refresher courses on a regular basis. There should be a complete "buy-in" to the system by all personnel at all levels of the company. It is pointless if department managers are recommending a system that those on the shop floor do not consider workable. If the people actually implementing the change do not conform to the system, there is, in fact, no system.

Once agreed on, the change control system should govern the entire site operations. If different departments operate their own independent systems, then the two words "change" and "control" become completely contradictory.

A change planned by one department may conflict with one already being considered by another. The same system should apply to operations performed on weekdays and holidays, by day and night shifts, by staff and contractors and by manufacturers and suppliers. If a company has various manufacturing sites, with personnel seconded from one to another to cope with particular projects or demands, then the same change control system should be in operation across each site.

All the relevant documentation to operate and maintain the facility should be provided as part of the construction project, in the form of technical manuals, and the change control system must ensure this documentation is updated wherever and whenever necessary.

Myriad changes will be made to a facility and its operations throughout its working life. Some will be planned, others will be knee-jerk reactions, quick remedies necessitated by system breakdowns. Some will improve matters, others will be detrimental. Some will be obvious, others deliberately less so. Some will be declared, others will never be discovered. All are potential pitfalls and all need to be accurately recorded.

Changes will be made by operators, maintenance personnel, contractors, service engineers, manufacturers, suppliers and consultants, among others. These people need to be controlled, and they need to be controlled in the least onerous manner. It cannot be emphasised enough that, if a change control system is difficult to work with, people will find ways of working around it. To make it as easy as possible for everybody, the system has to be simple, developed with the site working practices firmly in mind. It should complement, not contradict, other systems already operating on the site. If there is conflict, one of the systems will be on the losing side. If the loser is the change control system, there is no change control system. In particular, those responsible for effecting changes should be consulted to find the best approach. Physical changes to plant installations are performed, more often than not, by engineering or maintenance department fitters — people who traditionally work with tools, not pens. A system designed to accommodate their existing job sheets or work instructions is one most likely to succeed over another system requiring additional documentation.

Once effective, the system needs to be strictly enforced, with strong disciplinary measures imposed in instances of noncompliance.

V. Permit-to-Work System

All personnel based on site must be adequately instructed in the operation of the change control system. Also, all those responsible for the actions of outside contractors and suppliers need to ensure that the system is properly adhered to. Unsupervised strangers, unacquainted with the site and its manufacturing operations, are a danger to others as well as themselves. Nevertheless, constant supervision of visitors should not be necessary unless they are in a particularly sensitive area. Instead, there should be other ways of ensuring the appropriateness of their actions and behavior. It is absolutely vital that they are conscious of the possible impact they may have on product quality, operator safety, and the environment. More specifically relevant to the pharmaceutical industry, off-site personnel working on critical systems (i.e., those that may have an adverse effect on the quality of the manufactured product) should be made aware that any unplanned change or deviation from the agreed-on work content may, ultimately, put patients' health or lives at risk. They should be given a sufficiently detailed knowledge of the system to be able to assess the impact that any aspect of their work may have on its critical operating parameters.

A permit-to-work system should control the activities of all visitors whose work is likely to result in a change of any description. It should embrace the change control system, ensuring that all changes are captured immediately, so their impacts can be adequately assessed. All physical changes to installations should be organized and supervised by the engineering department. The permit-to-work system should be developed and managed by the engineering department in conjunction with the health and safety department as a vital control tool. Without such a system, incidents such as the following would occur frequently.

> The inquest confirmed the justification of the charge of gross negligence and the company was appropriately penalized — according to the letter of the law that is, but not in the opinion of the victim's pregnant wife and two young children.

> He had been working on a routine cleaning exercise and was opening the manway to the vessel, as he had so many times before, when suddenly, the lid blew open, throwing him across the room, tragically breaking his neck in the process.

> He had followed the standard operating procedure (SOP) almost to the letter; after all, he'd written it himself. The message displayed by the distributed control system (DCS) told him the vessel had been evacuated. What it did not tell him was that the DCS service engineer had paid the site a visit the previous day and had left them with the benefits of the latest version of software, installed and ready to use. The "benefits," in this particular instance, meant that the pressure in the vessel was now undetectable. He couldn't actually follow the SOP completely because the vessel's pressure gauge had been removed for annual calibration.

> The investigation led to the revelation that nobody on-site, other than the woman at reception, had been aware of the service engineer's attendance. Recognizing the visitor, she had recorded the relevant details in the visitor's book and obtained his signature. She had not been able to contact the plant engineering manager on the phone, probably because it was lunchtime, but the service engineer knew his way around the site and would catch up with him later. He didn't, but his report would be

there within a week explaining what he'd done. How things had changed on site in that week....

Our poor operative was a victim of his own assumption. In his mind, all was normal. There was no reason for him to think anything was out of the ordinary. He should have followed the SOP and waited until the pressure gauge had been replaced, but who amongst us has not been guilty of a minor deviation from the written law? The pressure gauge was used only for secondary indication anyway; the first port of call was, and always had been, the DCS display. In light of this distressing example, it is not difficult to see why rapid communication of a change is a fundamental requirement of any change control system.

It is all too easy for a person to become overly familiar with a system he or she operates on a regular basis, but this familiarity, as the saying goes, breeds contempt. The pressure gauge had been installed for a very good reason: to prevent such incidents from occurring. But, unless this sort of information is imparted to operators, how are they supposed to fully appreciate the possible effect of such a change? If the system was deemed, at design stage, to require the "belt and braces" approach of dual pressure monitoring systems, this should have been translated into the SOPs for the system. The operator would then have been made aware of the importance of checking both displays during training in the procedures. Maybe then he would have waited for the pressure gauge to be reinstated. It is very easy for this sort of information to be overlooked in the writing of operating procedures, since they are generally written by operators without any engineering department input. The obvious way to avoid this happening is to ensure that engineering and health and safety department personnel, and, if possible, those who were consulted at the hazard and operability study (HAZOP) stage of the design process, are also involved in the review of operating procedures.

The following incident is another case of familiarity breeding contempt — this time on the part of an outside contractor:

There was no formal permit-to-work system on site, and the refrigeration engineer had been there many times in the past. He had always passed the time of day with the security guard at the gatehouse; he knew how to work the system to his advantage. When informed that the shift engineer had left the site for a moment, the contractor said he was going to be there for only about 10 minutes or so and would, in all likelihood, be off site before the shift engineer returned. The security guard, whose own shift was just about to finish, wished him all the best and let him through.

He climbed up onto the roof where the condensers were located and, determined not to get caught out again, pulled the ladder up onto the roof after him. When it was borrowed during his last visit, he had been stranded up there for nearly half an hour. He probably need not have bothered on this occasion because it was Saturday, and there were not many people in today.

It was precisely for this reason that the company had chosen that day for fumigating the clean room. How was he to know that the extract fan exhaust point, which was just upwind of the condensers, was going to start billowing out formaldehyde gas? Two hours later, the replacement security guard went to check if the refrigeration engineer was still on site. Not seeing his ladder, the guard presumed he had finished and gone. The other guard had obviously neglected to book him out.

They found the engineer's body on the Monday afternoon. If he had left his ladder up against the wall, the gassing supervisor may have spotted it before giving the all-clear to evacuate the gas, or the shift engineer may have set alarm bells ringing, but, then again, maybe not.

Permit-to-work systems have to encompass all the eventualities of allowing work to be executed on site by off-site personnel. The precise nature of the work and how it will be executed must be known and authorized by the engineering department in advance. Whether contractors intend to perform all activities using their own personnel, or subcontract part or all of the work, also needs to be known in advance. Authorization should not be given until all possible effects of the work have been formally assessed by appropriate representatives from the engineering, health and safety, quality, and production departments. Visitors should be made aware of all access routes, areas of restricted access, time constraints, and parallel operational activities and of the possible effects of any other relevant specific events.

A good policy to adopt when dealing with outside specialists is to request detailed method statements before any work is undertaken, no matter how simple the task may seem. The method statements should give precise details of the work to be carried out and what effect the work will have on any adjacent areas, connected systems, or equipment. All necessary safety precautions should also be included, as well as details of any hazardous materials to be used. These method statements do not just assist in understanding the full extent and impact of the work, enabling the appropriate preparations to be made, but they also serve as an aide-mémoire in checking that all changes have been fully recorded.

If, during the execution of their work, contractors recognize the need for a change to be made to the proposed and authorized scope of their activities, it is vital that they be instructed in what course of action to take. If they are working on a critical system, they should not continue unless authority is given by the on-site engineering department supervisor. If the system is noncritical, contractors may be allowed to make the necessary amendments to the agreed approach, as long as they give precise documented details of the deviation on completion of the work. For this process to function properly, it is essential that the contractor's idea of change concurs with the manufacturing company's definition.

The permit-to-work system should require all contractors, service engineers or consultants to inform the engineering department supervisor of any accidental damage to, breakage of, or interference with any facilities, utilities or equipment occasioned by their work. This should be done in such a way that the perpetrator is praised rather than punished for the revelation of such misdemeanors. The employment of a "please tell" policy should be actively encouraged in all operational environments so that any potential disasters caused by cover-ups or temporary fixes can be avoided. It is easy for the manufacturing company to operate such a policy for its own personnel, but it is not quite so easy to ensure that outside contractors apply the same amnesty to their work force, particularly if the damage has to be put right at the contractor's expense. A written guarantee of compliance with the policy should be a requirement for all service providers before any contractor is permitted to work on site.

The permit-to-work system must enable all changes, whether like-for-like replacements, breakdown remedials, or alterations, to be captured and assessed by the appropriate personnel so that any further actions can be managed by the site change control system. The engineering department should perform regular and detailed audits of the permit-to-work system to assess its robustness and should seek the assistance of visiting contractors to identify any areas of the system that need to be addressed.

VI. Traceability

Outside contractors responsible for the maintenance of a critical system must be strictly controlled so that a complete and traceable history of the system is ensured. For instance, if the valve seals in a pipework installation have to be manufactured from a particular material, the contract should specify this. If they have to be a particular type and grade of material and supplied by a particular manufacturer, the contractor should not be allowed any room for maneuver. In all probability, if there is scope in the contract for a cheaper, lower-quality alternative to be provided, then it will be.

It may be argued that the replacement of consumables, such as valve seals, does not constitute a change. Of course it does. Any form of replacement is a change. It may also be argued that, because a seal does not bear a serial number or unique identifier of any description, and it arrived in a plastic bag with 49 other similarly anonymous seals, then their traceability is impossible and it is pointless to try to tie the seal back to a specific purchase order. Not so. A particular valve type may have a number of optional seals, according to application, all of the same size and shape and, even possibly, identical colour. It would be very easy to grab a seal from the wrong bag and install it in an incompatible system. However, the operation of suitable traceability systems in parallel with, or part of, the change control system can prohibit, or at least minimize, the incidents of such mistakes. Even if a mistake is still made, it may be uncovered through a regular auditing process. In this instance, the delivery note accompanying the bag of seals should be checked against the purchase order. Once it has been confirmed that the correct product has been supplied, the bag of seals is stored in a designated area until required, clearly labeled for use only in the authorized applications. An SOP should ensure that, whenever a seal is removed from the bag, its destiny is formally documented, along with the signature of the remover and the removal date. The log book, or maintenance history, for the system in which the seal is finally installed should then ensure its traceability by documenting the seal's final installation position.

VII. Reasons For Change

At the beginning of this chapter, we were told to resist change wherever possible. However, in the real world, there are circumstances beyond our control that mean change is unavoidable and, more often than not, these changes fall into one of the following categories:

- Cost
- Availability
- Obsolescence
- Improvement
- Initiative
- Legislation
- Targets
- Reliability
- Automation

A. Cost

In this day and age, with so many companies seemingly run by accountants, cost plays an ever-increasing role in decision making. The search for more cost-effective alternatives can result in change, as can the realization that the purchase price of a particular system or item of equipment does not necessarily have any bearing on the cost of its upkeep. Although inexpensive to buy, it may be very expensive to maintain. Ever-increasing economic constraints, throughout every department of the organization, also take their toll on the steady state of a manufacturing company's operations — less handling and holding, quicker distribution and adoption of just-in-time principles and practices.

B. Availability

Nonavailability of approved parts is a frequently cited reason for the change to an alternative design or supplier. This is one of the worst types of change because it is not a system enhancer; it is unplanned and brought about by necessity. The way to prevent this happening is for a company not to have all of its eggs in one basket. A suitable stock of all manufacturers' recommended change and spare parts should be kept on site and, where possible, a single source of supply should be avoided. Guarantees should be sought from at least two suppliers that a stock of standard parts will be available at all times. These guarantees should not be taken as read; the suppliers should be audited to verify that the internal systems of control exist to ensure the necessary availability at short notice. For certain component parts, it may be advisable to set up contracts with suppliers over a long period to ensure availability. The cost of these backup services must be balanced or weighed against the cost of the parts not being available at all.

C. Obsolescence

As the plant gets older, changes are made purely to maintain levels of production. This is particularly applicable in the case of computer system hardware. The computer industry has a cavalier attitude toward hardware, with newer, improved versions constantly being made available. The ongoing support and backup for obsolete hardware becomes more and more difficult to maintain.

D. Improvement

Improvements to a process are inevitable, as demands for newer or higher-specification materials and products have to be met. The maintenance of that competitive edge, as the marketing people are always reminding us, is up to all of us. That means constant improvements are sought — an aesthetic tweak here, a package colour change there. It all contributes to the constant state of flux that is manufacturing in the late 21st century.

E. Initiative

As mentioned earlier; initiative goes hand in hand with a well-motivated work force. Potential problems are foreseen and obliterated; minor alterations are made to make everybody's life that little bit easier. It all helps add to job satisfaction. This is all well and good, and should not be discouraged, but should be permitted only in situations in which adequate systems exist to harness these changes at the times they occur.

F. Legislation

The new style of industrial law puts more emphasis on risk control and management, rather than presenting a set of hard and fast rules to abide by. Risk analysis is a continuous process and will constantly generate change. The first word of the term "current Good Manufacturing Practice" (cGMP) tells us that manufacturing operations and environments are always under the microscope of the regulatory authorities and, therefore, subject to change as new requirements find their way into the latest regulations and guidelines.

G. Targets

Each department within the company tightens its belt as commercial pressure is brought to bear, presenting with it yet another set of targets, all of which have to be achieved by further waste reduction and increased throughput. These aims can rarely be realized without considerable changes being made.

H. Reliability

Equipment is modified to try to minimize downtime. More robust components and longer-lasting change parts are constantly being reviewed in an attempt to improve reliability and plant availability.

I. Automation

As we strive for greater efficiency and overall economy, processes are made quicker and more user-friendly by the enhancement of their automatic control systems. Hardware and software are updated. Hardware is upgraded to ensure that the systems do not become obsolete and can be adequately maintained, and to incorporate additional disk space or new memory chips. Software is revised to adapt to changes in the manufacturing process or environment, to enhance the system with additional functionality, to correct errors in an earlier version, or simply as a recommendation of an equipment manufacturer or supplier.

   With regard to software, care should be taken to ensure that all the necessary functional tests are performed and accurately documented before the new version is accepted. The level of compliance with the latest guidance relating to electronic records and signatures may need to be assessed and verified. Revalidation of all critical control systems is essential, and all existing documentation needs to be updated to reflect any changes in version numbering, as well as access and password controls and backup copy storage requirements. One particular area of close examination should be the impact the latest version of an equipment supplier's software has on any macro programs written by the manufacturing company.

VIII. Documentation Changes

Following a change to an installation, all documentation affected by the change must be updated accordingly. There must be only one set of reference documents depicting the current state of an installation. All copies of the previously issued versions of the amended documents need to be collected, logged, and destroyed. Two different versions of the same document can lead to confusion. If an out-of-date document is still in circulation without any obvious indication that it has been superseded, it will be presumed to be the latest version and may be used to the detriment of the manufacturing process. The change control system should indicate the mechanism by which all recipients of individually numbered controlled copies of issued documents are informed of the update. It should also ensure that all superseded versions are accounted for. A missive should be sent to all department heads to announce any version changes in order for them to arrange the destruction of any uncontrolled copies, in instances where there are no distribution records kept. The issue and recall of controlled documents, even when they are specific engineering or production personnel, should be managed centrally by the quality department.

It is extremely difficult to envisage the knock-on effect of a change to a document without the use of a database. For instance, a simple change to a document number may take only seconds to execute, but what impact the change has on other documents may never be fully appreciated. There may be dozens of SOPs, manuals, drawings, work instructions, and other documents that cross-reference to the renumbered document, all of which need to be corrected as a consequence. They may have been drafted by a number of representatives from various departments with no individual person knowing just how many documents are affected by the change. Even if all the appropriate personnel were made aware of the change, it would be highly improbable that all document authors or users would realize the full implications and identify the need to update all the relevant documents. Some will have been written years before. Their authors, even if they are still around, cannot possibly be expected to remember every cross-reference they included.

If, as a matter of site-wide procedure, a reference section was included in every document generated (or was added to every document provided by an outside organization), a database could be compiled of every cross-reference. The mere push of a button would provide the necessary information to update all relevant documents. This information should be maintained by the quality department and should be sought at the time the change is proposed, so that the full implication of making the change can be appreciated. There may be other options available, and an alternative, more economical way of imparting the necessary information may be a possibility.

A change to a manufacturing process may be as simple as the addition of a product-holding vessel to increase throughput, for example, a couple of lengths of stainless steel pipework, a few valves, a vessel, and an instrument or two. Viewed in isolation, this would seem a very uncomplicated and inexpensive project, but the knock-on effects of the change could be widespread. There may be more cost involved in updating the existing documentation than in the actual project. If this is hard to believe, just have a look at the following list of documents that may be affected:

- HAZOP studies
- System description
- Equipment specifications
- Engineering line diagram (ELD), schematic or piping and instrumentation diagram (P&ID)
- Isometric drawings
- Layout drawing
- Procurement documentation
- Control software program
- Material traceability certification
- Certificates of conformity
- Equipment list
- Valve list
- Pressure test documents
- Passivation documents
- Operating and maintenance manuals
- Maintenance program
- Calibration program
- Maintenance SOPs
- Calibration SOPs
- Operation SOPs
- Cleaning SOPs
- Asset register
- Change request
- Change order

And this is only the tip of the iceberg. There may be weeks of time and effort tied up in revalidating the revised manufacturing process. Some would argue that nothing has really changed, that the new installation is a mirror of the previous one, but without the necessary validation work, this is pure assumption and, as we discovered earlier, an assumption can be a very dangerous thing. Depending on the company's approach to validation, some or all of the following documents may need to be revisited and some, perhaps in abbreviated form, will need to be reexecuted:

- Validation plan
- Design compliance review protocol(s), execution and report
- Installation qualification protocol(s), execution and report
- Operation qualification protocol(s), execution and report
- Performance qualification protocol(s), execution and report
  - Cleaning validation
  - Product validation

The reason for the "s" after the word "protocol" is that there may be several protocols that are affected by the change. If there have been minor changes made to other previously validated systems (for example, to extend their services to the new vessel), these will also have to be revised and partially reexecuted. For example:

- Clean steam — for sanitizing the vessel
- Purified water — for rinsing the vessel
- Nitrogen — for product blanketing
- Sterile compressed air — for product transfer

The resources needed to perform all this work are not inconsiderable, which is why the full extent of the change has to be taken into account at the initial assessment stage. It may well be that the *real* cost of a proposed change outweighs any benefit.

A like-for-like change, such as replacing a system component for an identical item, must be clearly defined within the change control procedure. A 3.0-kW totally enclosed fan-cooled motor, with a running speed of 2700 rpm, a full-load current of 6.2 amps, and a shaft size of 35 mm can be replaced by an identical motor with the same frame size from the same supplier and most people would agree this is a like-for-like change. Even so, the change needs to be documented because it will affect the system maintenance program. Drawings will not need to be amended, nor will manuals need to reflect the change, but there may be documents that do need to be updated. If the system in which the motor is installed has been validated, in all likelihood the installation qualification documents will contain a record of the (no longer applicable) motor serial number.

IX. Auditing the Change Control System

The responsibility for audits generally falls into the hands of the quality department and, in the case of the change control system, quite rightly. After all, it is a quality system and the quality department will be expected to defend the company during a regulatory authority investigation. As pointed out earlier, without a robust change control system, a company does not have a controlled manufacturing process.

There can be nothing more embarrassing or infuriating for a company than being informed by a regulatory authority investigator that an operator does not follow the company's own operating procedure. It might be because the operator has found a more efficient or less onerous way of working, or it might be a question of interpretation of the instructions. Whatever the reason, it should never happen, but it so frequently does. Comparing actions with instructions is the only sure method of finding out whether written procedures are current and accurate.

While using the example of a production SOP, but appreciating that the principles apply to the operations of every department across the manufacturing site, let us examine how this sort of change can be captured.

First, every SOP should be given a review date, that is, a date by which the document should be verified as still applicable, or changed to reflect the current method of operation. If an operation has changed since the SOP was written, the reason the change was not reported at the time it became effective should be explored. It might be that the operator did not recognize the shift in working pattern as a change. But a number of slight shifts can, in time, amount to a radical change in operation. If this is the case, the change control system needs to be modified, possibly just slightly reworded, to ensure that this type of deviation does not go unnoticed in future. If an operator has a genuine need to change a method of operation, the change should be formally requested, reviewed, approved, and authorized, and the SOP amended, before the actual operating method is changed.

Second, every SOP should be audited. In most manufacturing organizations, a production SOP will be audited by production personnel, which would initially seem sensible, but this is not the best form of challenge. It will probably have been written by a member of the production department, but there will also be engineering, health and safety, and quality issues associated with both the document and the actual operation. This is why it was emphasized earlier that the SOP review process should include departments other than the one immediately affected.

A team comprising individuals from the engineering, health and safety, quality and production departments will provide a much more analytical approach to auditing, as the examination is carried out from the perspectives of the different disciplines, rather than being biased to one particular function. The production department representative might not see the relevance of an operator leaning over a machine to free a trapped product, but other members of the audit team might see other possible implications. The health and safety representative may see an obvious danger, such as the need for guarding to protect the operator from moving parts. The engineering representative may see a less obvious one; the outstretched arm of the operator may be in danger of triggering an optical sensor and initiating a reject mechanism, or engaging a limit switch, which disables a safety device. The quality representative may not see any of these but may identify a possible product contamination issue.

A representative from one department may read a different meaning into an ambiguous instruction than another representative from a different department. By observing the operator in action, an engineering person may see an obvious area of improvement to the manufacturing process that would not be apparent by observing the machine in isolation.

One thing that should be emphasized during an audit is the need for operators to carry out their tasks typically, that is, exactly as they normally do and not as the SOP tells them to. There should be no recriminations for deviations from the SOP (unless the procedure is being deliberately ignored), as these deviations may highlight possible shortcomings in the change control system.

X. Summary

Change is a constant factor in the operational phase of a manufacturing facility and there are many different contributors. Some of these are site based, some are external, but all have to be controlled. For this reason, there has to be in operation one rigidly enforced, sophisticated, yet uncomplicated change control system to encompass the whole of the site's operations.

5

The Change Control System

Peter W. Thomas

CONTENTS

I. Introduction

This chapter addresses the problems and pitfalls associated with devising and operating an engineering change control system and offers assistance in arriving at a workable solution. Flowcharts and generic forms are included to assist the reader in developing their own system.

II. Background

When reviewing this chapter for the second edition, I was initially expecting very little to have changed in the four or five years since writing it, but I was ignoring the very message in the last paragraph of the chapter — "change is an essential part of development and growth." The system I am to describe now is considerably enhanced in detail and scope over the previous system. In particular, the process now allows segregation of changes into different categories at an early stage to simplify the system, and the forms contain more information to prompt the user, thus ensuring higher quality information is recorded. However, the basic principles remain unchanged — identify, evaluate, authorize, and record. These are the cornerstones of the process and are repeated throughout the detailed procedures.

Previous chapters have dealt with the why of change control. Throughout history, in both technical and social contexts, uncontrolled change has resulted in catastrophic losses and many fatalities. Later chapters in this book deal with some of these instances in graphic detail if reminders are needed. Major losses in the chemical industry in the late 60s and 70s acted as a catalyst for the development of change control systems for modifications following the disciplines of a hazard and operability (HAZOP) study and the quantified approach of hazard analysis (HAZAN). These methods enabled the designer and operator to reduce the likelihood of failure occasioned by changes or modifications to an acceptable level, but because they tended to be cumbersome and time consuming, they were predominantly reserved for new projects or major modifications. A perception that prevails even today is that minor modifications are somehow less hazardous than major ones. An "o" ring in a mechanical seal on the shaft of a pump may be a low cost item, but changing it for the wrong elastomer could lead to a leak, the escape of flammable or toxic fluid, and major loss. In the pharmaceutical industry, it should be well accepted that even the smallest, insignificant cost modification may result in a major threat to patient safety and plant integrity, and even threaten the business if regulators perceive that uncontrolled change is taking place in a regulated facility.

For this reason, the change control system described in this chapter is meant to be applicable across the full range of changes from minor substitutions to major facility rebuilds.

III. Functions of the Change Control System

To work satisfactorily, the system has to perform the following functions:

- Capture the intent of the change.
- Evaluate the effects of the change.
- Authorize the change.
- Record the facts about the change and the new status.

These basic functions will have a number of stages, and different requirements, depending on the nature and scope of the change. The basic requirements for all changes will be consistent and the individual flowcharts for the various types of change will demonstrate the differences in detail. The following sections will describe the key functions in more detail.

A. Capture the Intent to Change

Capturing the intent to change is one of the most difficult steps to achieve, and yet it is so fundamental that if the system fails here, no matter how good subsequent parts are, the overall package will be useless.

At the outset, it is vitally important to enable people to recognize that they wish to make a change and then enable them to get a foot on the ladder of change without difficulty. Systems and procedures are often prepared with the regulator in mind to show that there is rigor and discipline in operational activities, perhaps without consideration for the persons expected to use and understand the system. Parallels of the legal profession spring to mind, where a lawyer is needed to read and interpret a contract. Training and use of any system helps, but if that system is inherently understandable and simple to initiate, compliance early on is more assured. In a manufacturing organization, the levels of management and their impact on change are as follows:

- Senior management — numerically few; may ask for change, but rarely actually make changes
- Quality departments — numerically larger; audit, monitor and to some extent control change, but again, do not actually make changes

- Engineering departments — frequently smaller than quality departments; prone to making changes during design and construction, but less so postvalidation

- Operation and maintenance departments — numerically large; on site four times as long as quality department, and potentially the greatest engine for change in operational plants

Clearly, having a complex and authoritative change control system may give senior management a feeling of comfort and quality departments a level of control. But if the operation and maintenance people are not fully aware of what constitutes a change, the validated status of the equipment and processes will rapidly decay, and compliance will be lost.

My company has made considerable efforts, through awareness training, to ensure that operational and maintenance staff are aware of change, know how to recognize it and how to initiate the process of requesting a change. We train using keywords, so if anyone raises a work request (the only vehicle for getting a job done), or receives a work request, with a keyword then a change is imminent. The keywords are:

CHANGE

ALTER

MODIFY

ADD

REMOVE

REPLACE "A" BY "B"

By training and example, we ensure that when a requestor prepares a work request containing a keyword or simile, or a tradesperson receives such a request, the need for further authorization is necessary. Triviality is not an excuse for circumventing the procedure, and no one is ever admonished for raising a request that is unnecessary.

The ability to recognize an impending change at the workplace is the key to success with the rest of the system. A further reinforcing strategy is to allow anyone in the organization to initiate a change request. The forms are freely available and simple to use. The form allows the initiator to describe the change and justify it. Further steps are necessary before the change takes place, but the originator can monitor the progress of the change when the change is accepted.

This ability to request changes increases an individual's feeling of empowerment and encourages compliance.

B. Evaluate the Effects of the Change

Once the request is made, it is submitted to a line manager for approval to proceed with evaluation. Clearly, owners need to approve what is happening

	Equivalence	Design Improvement	Minor Project
Initiate Change request			
Evaluate			
Authorize			
Record			

FIGURE 5.1
The matrix.

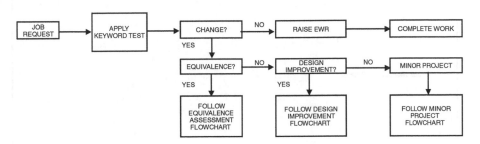

FIGURE 5.2
Decision tree for change.

to the plant or processes under their control. This also allows additional supporting information to be provided, or for refinement to be added before the design process begins. The line manager may alternatively feel that this is nice to have, rather than an essential improvement, and decide not to proceed. Requiring the originator to justify the change and the line manager to approve the principle of the change goes a long way toward weeding out unsustainable or unwarranted modifications.

At this stage it is necessary to introduce the vertical divisions in the system that, in conjunction with the horizontal steps above, form a matrix for handling change (Figure 5.1).

Changes can be divided into three broad categories — equivalence, design improvement, and minor project.

To enable the system to be tailored more specifically to the change, it is necessary to define the nature of the change in a way that leads the originator and designers down the correct route. The decision tree flowchart (Figure 5.2) shows the route through the stages. The change request forms are shown in Figure 5.3.

At this point, you might expect an increasing level of complexity, with a four-by-three matrix of operations, and the potential for a dozen different procedures. Dividing the process vertically, at this point, actually simplifies things. The steps involved in evaluating the effect of the change are driven by the type of change, but follow similar paths. Investigations will be made into the practicality, cost, quality impact, benefit and safety issues concerning the change. The owner will indicate his or her willingness to accept the change and other disciplines will become involved in design, costing, evaluation and approval.

ENGINEERING CHANGE REQUEST | ECR No. |

| EQUIPMENT DESIGNATION |

| EQUIPMENT LOCATION | EQUIPMENT No. |

REASON FOR CHANGE

Equipment improvement/replacement   ☐   New equipment/process   ☐

Quality/regulatory   ☐   Safety/Health/Environmental   ☐

Description of Change

Justification for Change

| Requested Completion Date |

Requestor Name   Signature   Date

OWNING DEPARTMENT AUTHORIZATION

TITLE   Signature   Date

ACCEPTANCE FOR DESIGN

APPROVED ☐   REJECTED ☐   AMENDED ☐

AMENDMENTS/REASON FOR REJECTION

ASSIGNED ENGINEER

ENGINEERING AUTHORIZATION   Date

SHE ☐   QA ☐   VALIDATION ☐   RA ☐

APPROVED NO FURTHER ACTION REQUIRED   ☐

APPROVED FOR DESIGN - DESIGN APPROVAL REQUIRED   ☐

HOLD PENDING FURTHER INFORMATION   ☐

Reason for hold

TITLE   Signature   Date

FIGURE 5.3
Engineering change request.

At this point, it is easy for the process to get weighed down with overheads, and to become cumbersome and slow. It is also the point where downgrading can artificially reduce the likely impact of a change by a desire to get something done quickly — haste rarely makes a good companion of judgement. The experience of other change control systems has shown that this is the stage where the effect of ensuring appropriate levels of approval, and adequate consideration by all concerned, has led to complex, multipage documents requiring signatories from all manner of departments. With this kind of escalating complexity, it is not unusual to find the number of change requests dwindling as the system backs up with unfinished changes and the almost certain result that unauthorized changes are happening in the field. This area is most difficult to regulate, and the author of a change control system will be under great pressure to provide more and more detail to cope with the increasing variants of change. At all costs try to maintain as generic a system as possible and allow flexibility within the system to cope with unusual situations. The broadening of our system to include three categories was made after a lot of investigation and consultation and only as a response to keeping the more basic change evaluations simple.

The evaluation process involves alerting other departments about impending change. By this method we can ensure that quality assurance, validation, regulatory affairs, health and safety, and environmental issues are dealt with adequately. This early alert to the other departments avoids the potentially wasted effort by designers and others if the system under review is part of a regulatory filing, and preapproval could be required. The other departments need to respond to the change notice and, by their positive response, will be included in the design approval process.

The request now moves to the design and specification phase, where it comes under the normal business control methods of estimation, cost control, financial authorization, and implementation. However, the actual change is still only at the appraisal stage; authorization is yet to come.

C. Authorizing the Change

The authorization process will broadly follow the same lines, regardless of the size of change. What will differ is the number of participants at the design review and the management level within the organization. The steps involve a design review, a HAZOP review and a current Good Manufacturing Practice (cGMP) review. The financial and business systems involved are outside the scope of this procedure and are not considered.

The preference is for the design review and HAZOP review to be conducted at the same meeting. It is very difficult to get a satisfactory response by sending the forms around to be signed off because the value of the review is reduced, and the time spent on successive desks — or lost in the mail — is likely to bring the process to a complete halt. The design and HAZOP reviews should be complete before an abbreviated meeting considers the GMP aspects of the change.

At these reviews, it is important to consider just the change and its impact; there can often be a tendency to go beyond the change and start redesigning or reviewing the whole plant. A strong chairperson — preferably not the originator or the designer — is needed to control these meetings. Occasionally, the review process will create a problem that cannot be accommodated

within the design. At this stage, stop the process. Do not be tempted to try to redesign the change at the meeting. Go back to the design phase, and reconsider. For a minor project involving a large number of components and some complexity, the main part of the process can be approved with a ballooned sector held for reassessment. The subsequent revision can be the subject of a design improvement within the scope of the original change.

D. Recording the Change and the New Status

After the design approval, the change is made precisely in accordance with the approved design. If it is necessary to deviate from the approved design for any reason, an equivalence form (Figure 5.4) or a design improvement form (Figure 5.5) will need to be completed and authorized. At this point, the designer and originator are interested in getting the physical change completed and moving on, and yet there is still much to do. The revalidation process needs to be completed, standard operating procedures need to be updated to reflect the change, and plant maintenance manuals may need to be changed. There is the need to update as-built drawings and equipment history files. There is also a need to carry out an on-site hazard study to ensure that the change has not given rise to unforeseen hazards at the plant. Safety checks concerning earth continuity or electrical loop testing may be required and will need to be recorded. Pressure systems will require testing and certifying, as will lifting equipment. Critical instrumentation will have to be recalibrated and all the documentation needs to be collected,

ENGINEERING EQUIVALENCE ASSESSMENT					ECR No.	

PART/ITEM DESCRIPTION		
DUTY/APPLICATION		
HOST EQUIPMENT REFERENCE		LOCATION

ELD Ref	Rev	Date	Tag No.

JUSTIFICATION	QUALITY	COST	OPERABILITY	AVAILABILITY	OTHER	

Validation Reference	IQ	Date	OQ	Date

IS THIS A PRODUCT CONTACT COMPONENT?	

DETAILS FOR REFERENCE	EXISTING	PROPOSED	NOTES
Manufacturer			
Supplier			
Part no./Ref			
Type			
Serial No.			
Other details			

DETAILS FOR ASSESSMENT		EXISTING	PROPOSED	NOTES
Key Attribute (size,range, duty)				
Product contact materials List all including seals and gaskets				
Pressure	Operating			
	Design			
Temperature	Operating			
	Design			
Other Parameter	Operating			
	Design			
Electrical Classification				
Unique component or new standard				
Other details				

Assessment prepared by Name	Sign	Date

Assessment approved by Name	Sign	Date

QA/Validation approval Name	Sign	Date

Further Action/Comments		
Name	Sign	Date

FIGURE 5.4
Engineering equivalence assessment.

collated, and stored in a coherent and secure manner. Not all changes will need to go through all stages. On minor projects this documentation can be considerable, but when substituting one item for another, much less will be required. To aid the process, we have included generic statements on our change control forms to provide prompts for the user.

The final step is to record, in a dedicated change request log, that the change is complete and where the documentation may be found. The change request log serves a useful purpose during and after the change process. It is basically a document tracking system and will record the change request number, the date raised and by whom, the assigned engineer, and the key dates for milestone activities like HAZOP and design reviews, and will

DESIGN IMPROVEMENT FORM	ECR No.		

ELD REF		REV No.	DATE
GA REFERENCE		REV No.	DATE
EQUIPMENT SPECIFICATION		REV No.	DATE
PROJECT REFERENCE AND TITLE			

CONDITION	YES	NO	CONDITION	YES	NO
Project design change			Replacement involving upgrade		
Commissioning modification			Nonprocess change		
Installation change in operation			Nonvalidated equipment change		
Break into site services			Environmental improvement		
Break into water system			Maintenance improvement		
Quality improvement			Other - describe		
Cost reduction/efficiency					
Equipment replacement					

CHANGE PROPOSED AND JUSTIFICATION

BUDGET COST ESTIMATE
TIME TO COMPLETE

DETAILED DESIGN INFORMATION TO BE ATTACHED

Prepared by Name	Sign	Date

APPROVALS

Engineering Name	Sign	Date

Owning Department Name	Sign	Date

FIGURE 5.5
Design improvement form.

indicate the current status of the change. During the lifetime of an individual change, the change request log is invaluable in allowing the change to be expedited, and after completion, it is a record of the change history and location of the supporting documentation.

IV. Types of Change

A. Equivalence

This is typically the result of obsolescence, resulting in the unavailability of a particular component or the desire to substitute one item with another for economic, performance or standardization reasons. Examples could include

Equivalence

4 Wheels ✓
4 Seats ✓   OK?
4 Cylinders ✓

an alternative hose manufacturer, a different mechanical seal supplier whose product is more easily sourced than the original equipment supplier's component, or the desire to add another valve manufacturer to the list of approved suppliers of a particular valve type. The process used here is to demonstrate equivalence by comparing key attributes, and for a competent person to authorize the change.

Design reviews and HAZOP studies are confined specifically to the substitution and its effect because one is comparing the items, not designing an application or changing operating conditions. A sample equivalence form is shown in Figure 5.4.

If a person is competent to carry out an equivalence assessment, they will complete the form and, having had the form approved by a second person of equal competence, will carry out the design review after an engineering change request (ECR) number is assigned. If the substitution is specific to one item of equipment, only then will the form be included in the equipment history file for that item and any validation data updated. If it affects a company standard, the standard will be updated referencing the ECR number.

B. Design Improvement

The design improvement is frequently encountered during a minor or major project and less frequently during plant operation. In the project context, design improvement can be the result of the integration of systems requiring a component change for compatibility reasons During commissioning, a particular configuration may not work well and it may be necessary to reconfigure to remove a problem. At the late stages of procurement, an item may have to be substituted for some reason (cost or availability), and the replacement may not be equivalent. A process review may indicate that additional functionality is required at a late stage in the project post-design freeze. All of these are instances of design improvements, generally minor changes, and low-cost, medium-impact modifications. They can be raised as a result of an ECR or can be raised directly by an approved designer. In both cases, the approved designer describes the change and justification with an additional tick box selection for the various reasons for change, and the effects on the equipment. The change information is all recorded on the design improvement form (Figure 5.5). The actual proposed design documents are appended and the package is signed off by a second competent person.

Design approval and HAZOP reviews are carried out. After authorization, the change is implemented. For a design improvement made during a project before handover, the completed package becomes part of the project documentation, and associated document packages are updated during the commissioning and validation processes. For design improvements on operating plants, the change is subject to the extended handover and validation activities that follow engineering work.

C. Minor Project

This category covers a large proportion of changes, generally those requiring modification to plants and equipment, with new components added and modifications made to operating procedures and process conditions. For changes of this type, there will be purchase orders involved, downtime at plants, and usually some kind of financial authorization involved. The process is basically the same, but the contents of each step will be greater than for the design improvement. The request will pass through the owning department approval prior to being assessed by engineering. A preliminary evaluation will show that the change is more extensive and will require financial authorization, so a preapproval stage will decide if the change adds sufficient value. Assuming a positive outcome, the request passes onto the designers for working up as a detailed and cost-estimated specification. Design and HAZOP reviews are conducted and, if necessary, a GMP review is carried out. The results are recorded on the plant design approval and HAZOP form (Figure 5.6). A design checklist (Figure 5.7) is included as part of the ECR package to prompt the assigned engineer to consider the requirements for additional information in support of the change.

Following approval and installation, the plant is recommissioned, additional validation is completed if required, and all associated documentation is updated as prompted on the plant handover certificate (Figure 5.8) and design checklist. Uncompleted actions or discrepancies are recorded on the back of the handover certificate to allow process testing and validation to continue. Finally, all documentation associated with the project is collated and transferred to the equipment history file for validated plants, or to the owner and keeper for unvalidated systems. The minor project process is detailed in Figure 5.9.

PLANT DESIGN APPROVAL AND HAZOP | ECR No.

TYPE OF CHANGE	EQUIVALENCE	DESIGN IMPROVEMENT		MINOR PROJECT	
DESCRIPTION					
HOST EQUIPMENT/TAG NO.					
LOCATION					

REVIEW CONSIDERATIONS				
SAFETY REVIEW	**IMPLICATIONS**		**COMMENTS**	
	YES	NO		
PRESSURE				
TEMPERATURE DEVIATION				
FLOW				
LEVEL/VOLUME & CONTAINMENT				
ELECTRICAL CLASSIFICATION				
POWER FAILURE				
AIR FAILURE				
COOLANT FAILURE				
OPERATION WITHIN DESIGN LIMITS				
ENVIRONMENTAL				
OTHER (specify)				
QUALITY REVIEW	**IMPLICATIONS**		**COMMENTS**	
	YES	NO		
COMPATIBILITY OF MATERIALS				
CHEMICAL RESISTANCE				
cGMP				
GAMP				
21CFR PART 11				
WATER SYSTEMS				
STATUTORY & OTHER	**IMPLICATIONS**		**COMMENTS**	
ISSUES REVIEW	YES	NO		
PRESSURE SYSTEMS				
LIFTING SYSTEMS				
BUILDING REGULATIONS (CODE)				
WATER HYGIENE				
OPERABILITY AND ERGONOMICS				
WASTE/ENVIRONMENTAL				

Assessor Name   Sign   Date

APPROVALS

Requestor   Name   Date

Engineering   Name   Date

Owning Department   Name   Date

SHE   Name   Date

R&D   Name   Date

Validation   Name   Date

QA   Name   Date

RA   Name   Date

Further Action/Comments

FIGURE 5.6
Plant design approval and HAZOP.

DESIGN CHECKLIST			ECR No.	

TYPE OF CHANGE	EQUIVALENCE		DESIGN IMPROVEMENT		MINOR PROJECT	
DESCRIPTION						
HOST EQUIPMENT/TAG NO.						
LOCATION						

REVIEW CONSIDERATIONS				
GENERAL DESIGN	**REQUIRED?** YES	NO	**NOTES/REFERENCES**	**INITIAL**
ATEX REVIEW				
CALCULATIONS				
CALIBRATION LIST				
CAPITAL EXPENDITURE PROPOSAL				
COMPANY STANDARDS				
CRITICAL PARAMETER LIST				
DRAINAGE				
ELD				
ELECTRICAL AREA CLASSIFICATION				
ELECTRICAL LOOP DRAWINGS				
EMERGENCY LIGHTING				
EMISSIONS CALCULATIONS				
EQUIPMENT LIST				
EQUIPMENT SPECIFICATION SHEETS				
ERGONOMIC REVIEW				
HEAT BALANCES				
I/O LISTING				
INSTRUMENT LIST				
INSTRUMENT LOOP DRAWINGS				
LIFTING EQUIPMENT SCHEDULE				
LIST OF SUPPLIERS				
MAINTENANCE DOCUMENTS				
MASS BALANCES				
MATERIALS DATA SHEETS				
MEANS OF ESCAPE				
PRESSURE VESSEL LIST				
PROCESS DESCRIPTION				
PROCESS ENGINEERING DATA				
SCOPE OF WORK				
STATUTORY REGULATIONS				
STRUCTURAL CALCULATIONS				
UTILITIES LOADS				
VALVE SCHEDULE				
VALIDATION REQUIREMENTS	**REQUIRED?** YES	NO	**NOTES**	**INITIAL**
URS				
DQ				
IQ				
OQ				
PROCESS VALIDATION				

Assigned Engineer Name	Sign	Date

Checked by Name	Sign	Date

FIGURE 5.7
Design checklist.

V. Pitfalls and Problems

The one-size-fits-all approach is not going to work every time, and there are instances where the generic process does not suit the circumstances. In times like these, there is a tendency to write in extra details to cover that issue,

PLANT HANDOVER CERTIFICATE | **ECR No.**

ECR DESCRIPTION

EQUIPMENT REFERENCE | **LOCATION**

ELD Ref | Rev | Date

Plant Operating Procedure No. | Rev | Date

Design Checklist attached? Yes/No

Sign | Date

Is this a Temporary Change? Yes/No

Sign | Date

Is this a reversal of a Temporary Change? Yes/No

Sign | Date

INSTALLATION COMPLETED

Engineering | Name | Sign | Date

POST-INSTALLATION HAZOP COMPLETED

SHE | Name | Sign | Date

Engineering | Name | Sign | Date

SAFETY CHECKS COMPLETED-DESCRIBE

Engineering | Name | Sign | Date

ACCEPTED BY

Owning Department | Name | Sign | Date

VALIDATION COMPLETED

Validation | Name | Sign | Date

QA | Name | Sign | Date

Further Action/Comments

Name | Sign | Date

FIGURE 5.8
Plant handover certificate.

should it arise again. The next issue is slightly different, so more details are added and so on until the system has grown to become the lumbering beast that threatens to stifle development and progress toward more efficient plants.

This creep of detail is difficult to resist, but worth doing. If possible, do not ask what you should add to ensure that you cover this eventuality, but instead ask if you are able to make the system less prescriptive, and therefore likely to include more unusual eventualities.

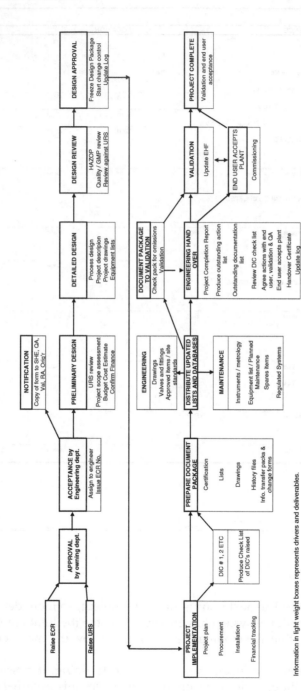

Information in light weight boxes represents drivers and deliverables.

FIGURE 5.9

Minor project — flowchart.

Some of the more common problems are discussed next, along with ideas of how the standard system will address them.

A. The Temporary Change

Since the proposed change will only be in place for 24 hours while the correct item is repaired, certainly there is no need to go through a temporary change procedure. Long before the paperwork is done, the plant will be back to normal. The world is littered with the debris of the outcomes of temporary change — with Flixborough (see Chapter 6, Section V.A), perhaps the most tragic.

Clearly, temporary changes need to be considered in as much detail as permanent ones. The documentation needs to be better in the case of a temporary substitution because after the plant is restored to normal the evidence is gone and can no longer be referred to.

The documentation may need to be produced quickly, and reviews may be done on the plant in front of the kit, but the review documentation must still be produced. Temporary operating procedures will be necessary and batch records may well need to carry a deviation note to say the temporary change is in operation. At some point, the change will be reversed to normal and another set of documentation will be required, confirming that the temporary change has been restored to normal. All this can be done using the standard forms and normal procedure; the handover certificate has a prompt for temporary changes and their restoration.

B. The Midnight Breakdown

Many pharmaceutical processes operate 24 hours a day and the midnight breakdown, with a potential loss of an expensive batch of material, can and will happen. It is not unusual, at this time, for the plant to be in the hands of one manager and several shift supervisors and technicians. The question of a review team does not arise, and probably only one person who would qualify as a competent reviewer is available.

If the plant was at risk or a major loss was imminent, most managers and engineers would respond to the out-of-hours call and get in the car to assist. But even if there were no responders, the process could still be started. The ECR could be raised and the first steps taken —is it equivalent, is it a design improvement, because it certainly is not a minor project at that time of night.

Details could be collected, and if the person on site feels confident enough to make the assessment, the prompts on the forms will guide them through the process. Immediate handover of the completed documentation to the oncoming day staff will allow the procedure to be checked, confirmed, and recorded. If the on-site, responsible person is not prepared to take the decision and is unable to gain assistance, the process will have to remain shut down. In these cases, authority and responsibility go together with confidence and knowledge, and unless all aspects are satisfactory, the default position must be to take the known safe route.

C. Control and Computer System Changes

Chapter 6 deals with computer system changes in detail. Generally, the normal change control system will act as the backbone of a computer change control system. The request is raised in the same way, a review of the change is made under the same conditions as the design review, and a HAZOP review is still required. Documentation and procedure updates follow the same general course and detail issues regarding GMP compliance, testing, version upgrade, security, and so on will be dealt with as additional packages.

D. What Systems Are Subject to Change Control

The fact that no one should be admonished for raising an unnecessary change request demonstrates our view that few changes will not need approval or recording. All process plants and utilities will need to go through change control; electrical systems, data, and communications will similarly need to be controlled. Indeed, it is difficult to think of many areas where control is not necessary, and, even if the change is trivial and not on a controlled system, the raising of a request will generate a level of involvement and scrutiny to ensure the change is appropriate and safe.

E. Assigning Authority Levels

What criteria are used to decide the signatories on change control documents? Who has the authority to approve a change of supplier or approve a whole new manufacturing line as being ready for construction? This is where difficulties occur. The only advice is to use the authorization limits that already apply to your organization. The process that selects and authorizes people to sign off documents or raise purchase orders is perfectly suited to signatories for change control purposes. All along, the message is simplicity, transparency, and keeping it as generic as possible. Avoid inventing a special authorization level for change control purposes.

VI. The Change Control Forms

The forms illustrated are reduced versions of the forms currently used by my company. The boilerplate content has been removed so that only the change control process information remains. There would normally be additional information relating to the SOP, author, approver, form reference number, company name, and so forth. To enable the forms to fit the page, some areas are reduced. The design process will lead to numerous other documents, some of which are referenced in the design checklist (Figure 5.7). However, all engineering change control packages should at least have the following:

- The change request
- The equivalence or design improvement form (unless a minor project)
- Design approval and HAZOP form
- Handover certificate

All the forms used to illustrate the system are paper copies, but electronic versions may be used. The prompt process can be much subtler with submenus and the like. The questions of security, validated software, and long-term data back-up need to be considered. There are several maintenance work packages that are capable of handling this kind of system, and they have been validated for maintenance and calibration purposes. They should be capable of extending to an electronic system for change control documentation.

A. Engineering Change Request (Figure 5.1)

This is the starting point for the process and is used by the widest group of people. The form is intended to lead the requestor through the process, and starts with the location, title, and description of the equipment or plant to be modified. The reason for change is narrowed down to four categories. Space is also provided for the description of the proposed change — and a justification. The target date for completion is also requested. This allows for the change to be given due priority later on in the process. At this point, the originator's part is complete and the form passes to the "owning" department supervisor or line manger. The change is assessed and the line manager's signature indicates acceptance of the need for the change.

The form then passes to the engineering department for scrutiny. (The first pass review categorizes it as equivalence, design improvement, or minor project, and the appropriate person is assigned to take the change forward.) The ECR is logged and given a unique number, and copies are made for distribution to Safety, Health, and Environment (SHE); Quality Assurance (QA); and validation and regulatory affairs. Copies also are made for the originator and for the engineering department's ECR file. Occasionally, it is apparent that the change will not be possible in its requested form, so amendments may be made at the outset and passed back to the originator.

Sometimes the change is impossible to progress and will be rejected. In this case, only the originator copy will be raised, and the ECR will not be assigned or given a number. The copies given to other departments will return to the assigned engineer in due course and will alert them to the level of sign-off required and the need for validation and so on.

B. Engineering Equivalence Assessment (Figure 5.4)

This form is used as a comparison tool so it is easy to see whether key critical parameters are the same. The general criteria for accepting that an item is equivalent are that these key attributes are the same or better. The concept of "better" needs to be defined by the person making the comparison in the context of that particular application. For instance, PTFE may have far greater tolerance to high temperatures and chemicals than nitrile rubber, but it makes an extremely poor O-ring due to its plasticity and lack of resilience. The application determines whether something is better and the person carrying out the comparison must be fully conversant with the application and the materials. The second-person sign-off is a standard checking process and broadens the experience relied upon during the approval process. The equivalence assessment process is mapped in Figure 5.10.

C. Design Improvement Form (Figure 5.5)

This form is used for changes that occur as a result of modifications incurred within the design and construction phase of a project, and for changes that do not require major modifications to equipment and systems. The form starts with the normal requirements for details of the equipment and location, and then gives a series of conditions that could apply to that change. All that apply should be highlighted. Once again, the justification for the change is required, as well as details of cost and time. These are particularly relevant in the case of a change being made on a site project. Signatures and approvals are required from the person preparing the design improvement form and from engineering and the owner, in the case of existing plant. The steps in the design improvement process are detailed in Figure 5.11.

D. Plant Design Approval and HAZOP Form (Figure 5.6)

All changes go through the design approval stage. For equivalence, it is a shorter process than for a minor project, with a design improvement falling between the two. The form contains the basic details of the change — and the location — at the top, and then, by a series of prompts, asks the reviewers to consider various safety, quality, and statutory factors.

The list is not exhaustive. More information will be required according to the standards already in place at your organization and according to the

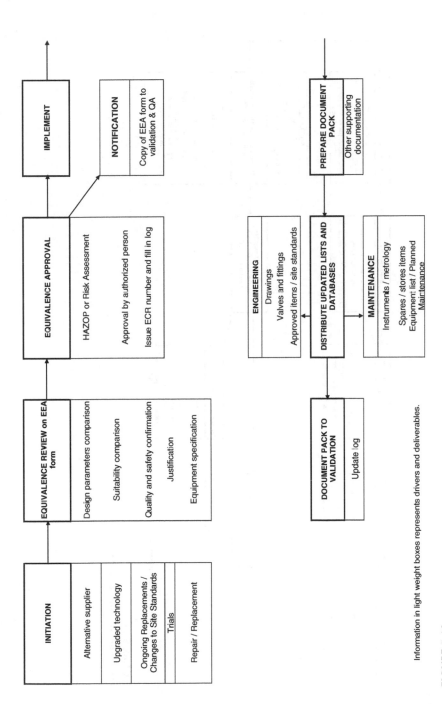

FIGURE 5.10
Equivalence assessment process.

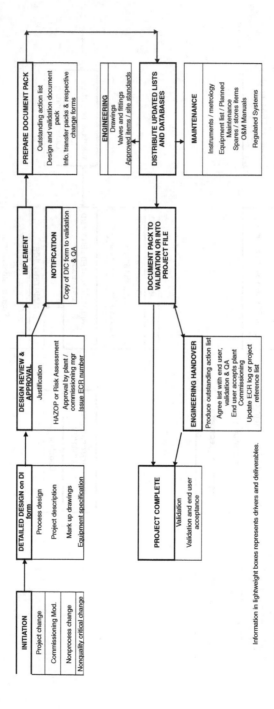

FIGURE 5.11
Design improvement process flowchart.

demands of the state. The list given is typical of the range that would be included at the review. Approval is required from the originator, owning department, and SHE departments. Other departments will be involved in approval, according to the nature of the change. Validated systems will include QA and validation as a matter of course and regulatory affairs will be interested in changes that concern filed processes. Research and technical departments will be required on changes that involve changes to the chemical process. The design approval stage appears in all steps.

E. Design Checklist (Figure 5.7)

This is an *aide-mémoire* to the designer and reviewers. It is a list of documents and stages that may form part of the change package. It is not exhaustive; again the list has been drawn up from typical documentation and activity requirements. Individual companies will certainly have more standard forms for information that would be referenced here. The intention is to convey the intent of this stage by giving example requirements.

F. Handover Certificate (Figure 5.8)

The final stage of the process is entered with the completion of this document. Cross-referencing the change, location and tag number with the design checklist attached is the point at which the installer, owner, safety, and quality departments come together to confirm acceptance that the work has been done. There is space for a few comments or deviations to allow handover to take place in the absence of some noncritical information. Once again all types of change employ this form at the conclusion of the change.

VII. Administration and Auditing

With any activity within an organization, it is imperative that someone has defined ownership, (see Chapter 7, Section XXVI), otherwise if left to many, each will assume the other is looking after it. With a process as important as change control in a pharmaceutical organization it needs to be part of an individual's primary job responsibilities. If this clear responsibility is not defined and accepted, the system will fall into disuse, abuse, and noncompliance, with disastrous effects at a regulatory inspection or during annual report time.

Even with clear responsibility for the system, audits will be required to ensure that ECRs are being closed out in a timely manner, documentation is stored where noted in the ECR log, and the quality of the decision-making process is compliant with current demands. Audit frequency is decided by the number of noncompliances and every organization will have a policy covering this aspect of the system.

VIII. Conclusions

The engineering change control system is a necessary part of your compliance armory. Without it, you will be unable to rely on the data stored in your validation and engineering documents. Even in a small organization that has only the most rudimentary, in-house engineering capability, there is a requirement to control the activities of contractors and outside servicing companies. To manage and operate the system requires commitment, vigilance, and a level of patience. The drawbacks are nothing when compared with the effects of not having a reliable system.

The flowcharts and forms will help mature organizations refine their processes and will enable emerging companies to start a viable system of their own.

Once again, I return to the concept that change is ever present and will occur. Ignore it at your peril. Resistance to change is futile; a fall in temperature of a few degrees will give drops of water the power to split mountains. Failure to embrace change in organizations results in mounting pressure and an inability to develop to meet the needs of the world at large. Control change and you will reap the benefits of evolution with confidence.

6

Change Control and Computer Systems

Tony J. Margetts

CONTENTS

I. Objective

Change control is fundamental to maintaining regulated products and to quality systems. In this chapter we will:

- Explain what change control is and how it is applied to computer systems.
- Describe some serious failures that have occurred as a result of not having good practices in place.
- Explain the links to good documentation and engineering practice.
- Provide details of typical change control systems.

II. Introduction: Change and Computers

Change is not new. Even the Greeks knew about change — they had a saying that nothing endures more than change. However, the pace of change has increased tremendously in the last 30 years as new technologies and new ways of working have been increasingly and rapidly adopted.

The amazing increase in the use of computers in industry has been one of the major features of change in this time period. Computers are now used for monitoring and controlling processes, products, people, and organizations. Not only do the processes, products, people, and organizations change, but the computers that control them also change. Suppliers change hardware and software with bewildering, and at times alarming, rapidity, and demands for changes to the processes, products, people, and organizations usually result in some form of software change because of the pervasive nature of the use of computers in the industry today. A particular problem with computer systems is that it is very easy to change the software.

III. Change Control — Basics

Before discussing the details of change control applied to computer systems, it is necessary to develop some basic concepts.

Change control is discussed in the ISO 9000 and 9004 procedures.[1-3] These procedures distinguish between design change control and process change control as follows:

Design change control — The emphasis is on procedures for controlling the release, change, and use of documents that define the design baseline

Process change control — The emphasis is on the authorization of process changes, which should be clearly designated, and if necessary, customer approval should be sought

ISO provides these two views of the change control process to differentiate between (a) companies that are design-based, where the product (in this case, a computer system) is supplied to a customer (in this case, a pharmaceutical company) and (b) companies that operate a process for making a product (in this case, the process involves the use of a computer system to make a pharmaceutical product).

In practice, a company should have one basic procedure to address their change requirements. The content of such a procedure for a typical pharmaceutical company will be developed in this chapter.

IV. Computer System Elements and the Operating Environment

Figure 6.1 shows a diagram of the elements that make up a computer system. The important elements are:

- The computer system
- The controlled process
- Operating procedures
- Controlled documentation

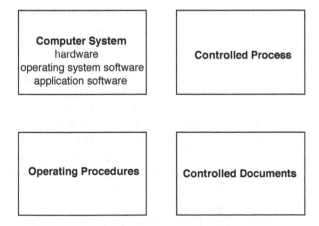

FIGURE 6.1
Computer systems and their operating environments.

The computer system generally consists of:

- Hardware
- Operating system software
- Application software

The computer system is defined by the assembly of specified hardware, operating system software, and application software, referenced by serial numbers and version numbers. The operating procedures are documents that describe the steps taken to run the process and operate the computer. These are referenced by titles, dates, and version numbers. The controlled process is the physical equipment and manual operations controlled or directed by the computer. The equipment is referenced by equipment reference numbers; interfaces are referenced by serial numbers. The process is described in a version-controlled document. Documentation is the collection of computer and equipment specifications, test specifications, test data, operating data, batch records, and training records, which must be controlled and retained for a defined period to support the operation of the plant and the production of the product.

The operating environment is the sum of the above elements and can be stated as the combined workflow between people and machines to accomplish the stated task. This is sometimes called a computerized or automated system. In the pharmaceutical industry, the complete environment, including the elements described above, is subject to documented change control according to the procedures defined in this chapter.

V. Considering the Consequences of Change

A. Some Serious Incidents That Have Occurred during the Use of Computer Systems

Serious incidents can occur if a change is made without a proper understanding of the potential consequences. One of the most serious consequences of lack of change control in the British chemical industry occurred at Flixborough in 1974.[4] A good summary of this most important case is given in the compilation of case histories toward the end of this book. Another serious issue that affects all of us is bovine spongiform encephalopathy, which has developed in the general population recently as a result of a change in the way animal waste was processed in the 1980s.

The increasing use of software to control equipment has lead to the occurrence of serious incidents directly attributable to errors in the software. Lack of change control has been a contributor to a number of these incidents.

1. Military Accident

A military accident occurred involving a prototype computer-controlled torpedo. The initial design was subject to a failure mode and effect analysis of software and the following possible failure mode was recognized. If the homing device failed after launch, there was a possibility for the torpedo to turn back on itself and to hit the vessel that launched it. The software was changed to detect a 180° turn after launch, in which case the torpedo would self-destruct. During trials, a torpedo was launched but jammed in the tube in the submarine. The captain ordered the boat to return to port, the boat swung around through 180° — bang!

2. Errors in Software Lead to Serious Loss

The Ariane 5 rocket was destroyed on its first flight as a result of a failure of the guidance system. Changes to the guidance system used in Ariane 4 had not taken into account the different trajectory of Ariane 5 because of its larger size.

3. Poor Software Can Kill

The Therac 25 was a computer-controlled hospital radiation machine. Leverson and Turner[5] describe how poor software engineering practice and lack of change control resulted in the deaths of six people in what has been described as the worst series of incidents in the history of radiation treatment.

B. Pharmaceutical Industry Failures

I have encountered the following faults in pharmaceutical computer systems:

1. Water for injection (WFI) system was jumping cycles when influenced by external radio frequency interference from a mobile radio.
2. Water for injection (WFI) purging routines incorrect.
3. Terminal sterilization autoclave had intended to have a warning message to say "sterilization temperature not achieved." When the program ran, the system triggered the message at the correct point, but the message text was blank and would not have been noticed by the operator.
4. Math error in an F(0) calculation in an autoclave.
5. System clock giving the wrong time in an autoclave control system.
6. Dry heat sterilizer oven display showed that the cycle had run and completed. The cycle had not run.

7. Storage tank overflowed when printer jammed. The jammed printer
had stopped the computer polling. Many apparently simple changes
can introduce errors such as those above, and it is essential that all
changes are controlled by a change process that involves a review
by knowledgeable, responsible people who will consider the consequence of the change.

C. Learning from Real-Life Incidents

1. Change Control Problems Involving Computers

Two executives were happily using their laptops while the SABENA
Airbus 340 was cruising. During the flight, both laptops began to behave
oddly, and gradually the hard disks failed. By the end of the flight, they
were both useless. The executives complained to the airline. SABENA
had been doing some customer research and discovered that the drop-
down trays in the seat backs rattled when stowed, so to stop the rattle
they had fitted new magnetic catches to the trays. These magnets caused
the laptops to fail. Europe's first Ariane 5 rocket exploded soon after
takeoff because its computer software, which had been transferred from
its Ariane 4 predecessor, could not cope with the bigger rocket and was
unable to control it. The design team admitted that, because of time
pressure, not all of the software had been checked and revalidated with
the new control parameters.

2. Difficulties of Testing

During the Falklands War, the HMS Sheffield was sunk by an exocet
missile. Interference from the ship's radio prevented the ship's warning
system from picking up signals from an approaching missile until too
late. All the systems had never been tested together.

During the first Gulf War, a scud missile got past the patriot antimissile
missiles. It was discovered that a computer clock at the missile outpost
battery ran slowly because of interruptions by a higher-priority task.
Patriots had never before been in continuous operation long enough for
the difference to be spotted.

3. Being Clear about the Main Protection Guarding Against the Risk

When the Lancashire & Yorkshire Railway built a dock at Fleetwood, the
quayside had to be sloped to meet local engineering constraints. The
operating department was told of the problems, and instructions were
issued requiring all wagons parked on the sidings to have their brakes
pinned down. Substantial buffer stops were fixed at the siding ends.
There were several runaways when shunters failed to set the brakes. In

every case, the buffers failed to hold, and the wagons ended up in the dock. The permanent way department got fed up with replacing the buffer stops and left the sidings without them. There were no more runaways.

4. Problems with Control Systems

A monorail was being installed for the 1990 Gateshead garden festival. The six trains were automatic and driverless, stopping at each station on the continuous loop, then starting again on a signal from the station controller. The detection system stopped a train when it got within a specified distance of the train in front. If the train behind stopped too close, it then reversed until it was far enough back A full test was set up and one of the trains overran because the rail was wet. It began to reverse. The train behind, finding a train reversing toward it also reversed. Within a minute all six trains were in reverse. They could only be stopped by cutting the power supply to the whole system. Because it was a test, there were no passengers on board at the time.

5. Computer Failures Can Be Share Price Sensitive

The failure shown in Figure 6.2 knocked £800 million off the share price of ICI.

FIGURE 6.2
Share price impact (From *Financial Mail*, March 30, 2003).

D. How to Identify the Consequences of the Change

Identifying hazards associated with a system and considering the consequences of particular actions require a structured approach. Study techniques have been described by Gillett.[6] For computer systems, a good technique is to conduct a study of possible threats to the correct operation of the system and the controls necessary to ensure that possible threats do not occur. This is called a threats and controls study and is a type of risk assessment, as described in Appendix 6.3. For hardware configurations, failure mode and effect analysis (FMEA) is appropriate, as described by Gillett. Any change to a computer system should be reviewed, and its consequences to the system should be carefully considered. The threats and controls method described in this chapter is a useful risk analysis technique. The change control procedure should prompt for such a study if appropriate.

VI. Change Control in Practice

This section considers how to turn the basics into practice. The first part considers some key ISO 900 quality system requirements for change control (Table 6.1). This leads to the introduction of good document practice (GDP) and good engineering practice (GEP), which are essential pillars of change control. The GDP and GEP concepts are more fully explained in the *Good Automated Manufacturing Practice* (GAMP) publication.[7]

VII. What Can Go Wrong

It is important to think about what can go wrong as part of the risk assessment process. Prevention is better than cure (Figure 6.3).

Appendix 6.1 gives an example procedure for change control suitable for use with computer systems operated in the pharmaceutical industry. It also brings out the points highlighted in Table 6.1, which emphasizes the importance of GEP and GDP. Both of these must be present in the organization before an effective change control system can operate.

FIGURE 6.3
What can go wrong?

TABLE 6.1

Key ISO 9004 Requirements and How These Can Be Applied to Computer Systems
in the Pharmaceutical Industry

Key ISO 9004 Requirements	How the Requirement Can Be Applied to Computer Systems
The presence of a quality management system	Companies that supply or operate a computer system in a pharmaceutical environment need to use good engineering practice
A procedure for controlling the release, change, and use of documents that define the design baseline; the configuration of the design baseline and its control process is referred to as configuration management	Computer systems require comprehensive documentation to describe the system in terms of requirements, design, installation, and operation; this documentation has to be maintained using good document practice
Those responsible for the authorization of process changes should be clearly designated; the procedure should provide for various necessary approvals	Authorization is required for the change to be started, and approval is needed to introduce the change when the change activities are complete
The procedure should institute formal design reviews and validation testing when the magnitude, complexity, or risk associated with the change warrants such actions	Use a threats and controls review for consequences and consider how much testing is required to support the change; often test plans used in the initial validation can be used again
Customer approval should be sought for product changes, where necessary	Key customers should be part of the change process for products
A product should be evaluated after any change to verify that the change instituted has had the desired effect upon the product quality	In the pharmaceutical industry this is covered by management and QA review of change processes and by annual product review
Change procedures should handle any emergency changes necessary	Prescribed emergency actions can be carried out by trained, knowledgeable and responsible persons, but in this case the change is raised retrospectively and approval is still required
Emergency changes may require new parts to be fitted	When a component has failed and a replacement part is fitted, which is a like-for-like replacement, this is not a change; such a replacement should be captured on a replacement log

Good document practice is: to ensure that key documents are created, reviewed, approved, distributed and stored in a controlled manner.

Good engineering practice is: the execution of engineering activities according to a formal quality management system that ensures proper handling of requirements, design, review, and audit of deliverables throughout the engineering (or change) life cycle.

A. GDP

Computer systems require comprehensive documentation to describe the system in terms of requirements, design, installation, and operation. This documentation has to be maintained using good document practice.

The purpose of good documentation practice is to ensure that key documents are created, reviewed, approved, distributed, and stored in a controlled manner. This ensures that key documents such as requirement specifications, design documents, test documents, and as-built documentation are used correctly. GDP is essential to establish a traceable and manageable basis for qualification and validation activities and as a basis for ongoing change control. GDP implies that:

- The documentation format should be standardized, including document layout, style, and reference numbering. Approval signatories should be identified by title in each document.

- Documents shall be under version control. The status of a version shall be identifiable (e.g., draft, approved, withdrawn, etc.).

- Prior to formal issue, a document should be in draft form. The version control method should reflect the document status of draft documents versus approved documents.

- Documents should be subjected to a formal review from which identified problems or proposals should be documented. Once all corrective actions identified during the review process have been completed the document should be updated and verified for correctness.

- Approval should be documented by signing, with formal signatures by two or more responsible people. There should be a statement of what the approval implies.

- Changes to the document should be approved accordingly. The formally reviewed copy of the draft document and review report should be retained. There may be several grades of reviews, in which case the type of review conducted should be documented. Review reports should be retained.

- Copies should be controlled and approved documents issued to controlled copy holders. Superseded documentation should be retained. Uncontrolled copies should be clearly marked "uncontrolled."

- A master document file should be maintained with the master document copy and a document history or change log. A master document index should be retained, showing document reference, document title, and document issue status.

- Superseded and withdrawn documents should be archived in a clearly labeled, separate file. Each document should be clearly marked as, for example, "withdrawn." If computerized distribution and archival is used, printouts should state that the paper copy is not a controlled version.

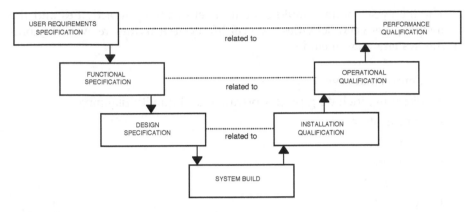

FIGURE 6.4
A basic validation and document life cycle.

A suitable validation and document life cycle is shown in Figure 6.4, which is from the International Society of Pharmaceutical Engineers (ISPE) GAMP guide.[7] This shows the different levels of specification related to the different levels of qualification. The same life cycle can be used for a change. The change has to be specified at the user level, again at the functional level, then at the design level. Once made, the change may be tested at the system level, at the design level, at the functional level, and then at the user level. Some or all of these levels may be used depending on the size and complexity of the change. An example procedure for the production, control, and issue of documentation that emphasizes the controls described above is given in Appendix 6.2.

B. GEP

GEP ensures the execution of engineering activities according to a formal quality management system, which ensures proper handling of requirements, design, review, and audit of deliverables throughout the engineering life cycle. Good engineering practice ensures that the engineering or software development methodology generates deliverables that support the requirements for qualification, validation, and change control in the pharmaceutical industry.

Good engineering practice implies a formal quality management system, which is based on an appropriate, well-defined project life cycle model (Figure 6.4), and includes support activities that are not life cycle dependent. The quality management system should clearly express the management's role and responsibilities and the quality policies of the company. It should ensure that activities are executed according to defined principles, and that quality records from the execution of activities are being generated and retained on a standard routine basis. GEP also implies the application of GDP, described above.

The life cycle model should contain a set of life cycle activities and the transitions between activities. Typical life cycle activities relevant to computer systems may include:

- Contractual review
- Planning, including project planning and quality planning
- Specification
- Design
- Construction
- Testing
- Acceptance
- Maintenance

The supporting activities should contain a set of activities that support all life cycle phases. Typical supporting activities relevant to computer systems may include:

- Reviews and inspections
- Document control
- Configuration management and change control
- Software control
- Control of subsuppliers and subcontracted work
- Control of incoming material including customer-supplied material
- Customer complaints
- Training and qualification of personnel

It is possible to control the documents using a computer, so that the procedure described in Appendix 6.2 can be followed by creating, approving, storing, and distributing documents on a computer system that is designed for this purpose. The approval process will involve the use of electronic signature. The change process can also be operated successfully on a computer system, and local and wide area networks can be used to communicate within the organization and with external suppliers and customers.

C. Configuration Management

Configuration management is closely related to the change control process. When changes are proposed, both change control and configuration management activities need to be considered in parallel, particularly when evaluating impacts of changes. Automated system components subject to configuration management include:

- Hardware (e.g., programmable logic controllers (PLCs), PCs, mini-computers, servers, communication interfaces, printers)
- Software code (e.g., PLC code, source code, executables, configuration files for configurable packages, data files, firmware, etc.)
- Third-party software (e.g., operating systems, library files, configurable packages, drivers, compilers); this includes software delivered with the system and customer supplied software items
- Manuals (e.g., user manuals, system manuals)

All components, and changes to them, should be controlled. The exact configuration of the system (hardware and software) should be documented throughout the life of the system. Further information is given in the ISPE GAMP guide.[7]

VIII. Auditing Computer Systems

Change control is a key topic during audits of computer systems — either by internal quality assurance staff or by external auditors, such as regulatory agencies when they audit for new drug applications and for compliance with manufacturing licences (or for certification to ISO or national standards).

Auditors can only check the status of a particular computer system using the controlled documentation and are likely to check the following:

- The existence of a formal change control procedure that is relevant and applies to the system being audited
- The existence of adequate documentation to show that the procedure is being operated effectively

Access control is another topic that interests auditors of computer systems. The main reason for their interest in access control is that unauthorized access can be a major cause of uncontrolled change, and there have been a number of instances of uncontrolled change as a result of management failure to control access. Auditors are particularly concerned about uncontrolled change to computer systems in the pharmaceutical industry. It seems to be a contradiction that software changes — which to a technologist not trained in GMP may seem easy to make — have to be made more difficult by written procedures. The GDP and GEP procedures, together with effective access control, provide the necessary defense against uncontrolled change to computer systems in our industry.

References

1. ISO 9000: (BS 5750) 2000, Quality Management Systems.
2. ISO 9004: (BS 5750), Part 0, Section 0.2, *Guide to Quality Management and Quality System Elements*, paragraph 8.8, Design Change Control.
3. ISO 9004: (BS 5750), Part 0, Section 0.2, *Guide to Quality Management and Quality System Elements*, paragraph 11.6, Process Change Control.
4. Gugan, K., *Unconfined Vapour Cloud Explosions*, IChemE, George Godwin, London, 1979.
5. Leverson, Nancy G. and Turner, Clark S. , An investigation of the Therac-25 accidents, IEEE Computer Journal, pp. 18 to 41, 1993.
6. Gillett, J.E., *Hazard Study in the Pharmaceutical Industry*, Interpharm Press, 1996.
7. GAMP4, *Good Automated Manufacturing Practice*, International Society of Pharmaceutical Engineers (ISPE), 2001.

Appendix 6.1

Example Procedure for Computer System Change Control

Processing Division
SOP 1-001v1
Product Manufacturing
Computer Systems Change Control

Written By: _____   Date: _____

Approved By: _____   Date: _____

Review Period: 3 years
Expiry Date: _____

1. Introduction

This procedure describes a system for the control of changes to automated systems. Change control is fundamental to maintaining validated processes and products.

2. Scope

This procedure applies to changes to controlled documentation, application software, operating system software, firmware, hardware and system configuration data, where referenced by a quality and product plan. The replacement of items on a like-for-like basis (e.g., following hard disk failure) may be controlled by other procedures defined in a maintenance plan rather than change control as described in this procedure.

3. Procedure

3.1 General

All changes shall be authorised, documented, tested and approved before implementation. However, the following changes are exempt:

- Emergency repairs
- Changes during approved experimental work

Any changes resulting from the above exemptions shall be subsequently reviewed, tested, documented, and approved in accordance with this procedure.

3.2 Request for Change

Any proposed change shall be requested by completing the request-for-change section of the change request form (Form 1). Each change request shall be assigned a unique reference number and logged using the change request index (Form 2).

3.3 Change Disposition and Authorization

Each system should have a designated project manager responsible for ensuring that all changes to the system are implemented in a controlled manner. The project manager may delegate this responsibility.

Each change request raised should be reviewed and its disposition (accept or reject) determined by management. At least two people are normally required to accept or reject a change. Items that have been approved by the pharmaceutical customer (e.g., contractual documents or tested and accepted software) should only be changed after prior approval of the change by the pharmaceutical customer.

3.3.1 Rejected Changes

Those responsible for rejecting the change request shall complete and sign the change disposition and authorization section of the change request, stating the reasons for rejection in the change details section of the change request. The change request should be filed, the index (Form 2) updated, and the originator informed of the decision.

3.3.2 Accepted Changes

Those responsible for authorizing the change request should complete and sign the change disposition and authorization section of the change request. They should also define:

- Which controlled items are affected by the change request
- The retesting required to ensure that satisfactory performance is maintained

These decisions shall be recorded in the change details section of the change request.

If more than one controlled item is to be changed as a result of the change request, a change note (Form 3) should be produced for each item, defining the particular changes to be made to that item. Note that some computer system changes can be complex and the use of change notes helps to control the complexity.

A change plan may be required for multiple activities and resources and, if produced, should be attached to the change request and cross-referenced from it.

3.4 Change Completion and Approval

When all changes have been implemented and retesting is complete the change request should be passed to management for final review and approval. The completed change request should be filed and the index (Form 2) updated.

Form 1	COMPUTER SYSTEM CHANGE REQUEST		
Change Number		**Date**	
Change Title			
Originator and person responsible for managing the change, phone, fax, e-mail			
Regulatory Action Required: Y/N	**Manufacturing Site Action Required: Y/N**	**Project Manager:**	
CHANGE DISPOSITION and AUTHORISATION Authorised/Rejected			

Signature/date/company

_____ _____

_____ _____

_____ _____

_____ _____

CHANGE DETAILS (Use change note for more detail; if necessary put comments include reasons for rejection or if accepted life cycle activities required)

Number of Change Notes attached	Threats and Controls or FMEA required
CHANGE COMPLETION AND FINAL APPROVAL FOR USE	
Project Manager	Product Manager
Date	Date
Regulatory Manager	Quality Assurance
Date	Date

Form 2	CHANGE REQUEST INDEX			
Change Number	Description	Disposition	Number of Change Notes	Date Closed

Form 3	CHANGE NOTE
Change Number	Change Note Number
Item Name	
Item Reference	Item Version Number
Detail of Change	

IMPLEMENTED BY
Signature   Print Name   Date

Appendix 6.2

Example Procedure for Computer System Document Control

Processing Division
SOP 1-002v1
Procedure for the Production, Control and Issue of
Documentation for Computer Systems

Written By: _____   Date: _____

Approved By: _____   Date: _____

Review Period: 3 years
Expiry Date: _____

1. Introduction

This procedure describes a system for the preparation, review, approval, and issue of documents.

2. Scope

This procedure applies to project documentation where referenced by a quality and project plan.

3. Procedure

3.1 Document Production

A procedure defining documentation standards should be agreed on locally. It should cover document layout, style and reference numbering. Documents should be produced in accordance with that procedure. Approval signatories should be identified by title in each document.

Prior to formal issue, a document should be in draft form, issue numbers should be identified alphabetically (e.g., starting with A), and then by a different mechanism (e.g., by whole number versions). Prior to formal review, the document is entirely the author's responsibility.

3.2 Document Review and Approval

Documents should be subjected to a formal review in accordance with a procedure for document review. When the review is held, a document history[1] should be opened. This history should be maintained in the master document file in accordance with Section 3.6 of this procedure.

Once all corrective actions identified during the review process have been completed, the document should be updated to the next level of issue, commencing with issue 1. Approval signatures, defined by the document or management, should be obtained using document approval.[3] Formal signatures are required by two or more responsible people. An approval signature should represent a commitment that the document has been thoroughly and completely reviewed by that person.

Further changes should be carried out in accordance with Section 3.5 of this procedure. The formally reviewed copy of the draft document and review report should be retained in accordance with Section 3.7 of this procedure.

3.3 Document Issue

Approved documents should be issued to controlled copy holders as defined by the document review minutes or management. The process followed to issue a document is as follows:

- Update the document history.[1]
- Update the master document index.[7]
- Open or update a document circulation register.[4]
- Issue a document transmittal notice[5] with each controlled copy.
- Write controlled copy number on document.

A copy of the document transmittal notices[5] should be retained until the signed original is returned. The above records should be maintained in the master document file in accordance with Section 3.6 of this procedure.

Superseded documentation should be retained in the archive document file in accordance with Section 3.7 of this procedure. Uncontrolled copies shall be clearly marked "uncontrolled." Changes to the document circulation register[4] should require prior agreement of management. Further controlled copies should be issued in accordance with bulleted items 3, 4 and 5 above.

3.4 Document Withdrawal

Documents should only be withdrawn upon receipt of a change request authorized by the approval signatories. The following should then be carried out:

- Update master document index[7] to indicate document issue status is withdrawn.
- Update document history.[1]
- Issue transmittal notices[5] according to document circulation register[4] such that all copies are destroyed by the holders.
- Archive documentation in accordance with Section 3.7 of this procedure.

3.5 Document Changes

Modifications to documents should be progressed in accordance with the procedure for change control.[2] The document should be updated to the next level of issue and reissued in accordance with Section 3.3 of this procedure.

Major modifications or rewriting of a document should be progressed as follows:

- Reset document issue to draft form (e.g., Draft 3A).
- Progress in accordance with Sections 3.1, 3.2 and 3.3 of this procedure.

3.6 Master Document File

A master document file should be maintained. The following should be retained for each document:

- The master document copy
- Document history[1]

For issued documents, the following should also be retained:

- Document approval[3] or change request(s)[8]
- Document circulation register[4]
- Document transmittal notices[5]

A master document index[7] should be retained, showing:

- Document reference
- Document title
- Document issue status

3.7 Archive Document File

Superseded and withdrawn documents should be archived in a clearly labeled, separate file. Each document should be clearly marked as "superseded" or "withdrawn." Review reports should be retained for draft documents.

The following should also be retained for each formal issue of a document:

- Document approval[3] or change request(s)[8]
- Document circulation register[4]
- Document transmittal notices[5]

An index of archived documents should be retained using the archived document index.[6]

References

A series of forms are required to record the necessary information. All references, except 2, refer to a suitable form for recording the necessary information. Forms can be easily prepared using Word tables.

1. Document history
2. Procedure for change control

3. Document approval
4. Document circulation register
5. Document transmittal notice
6. Archived document index
7. Master document index
8. Change request

Appendix 6.3

Example Procedure for Computer System Threats and Controls Review

Processing Division
SOP 1-003v1
Procedure for Threats and Controls Review
for Computer Systems

Written By: _____   Date: _____

Approved By: _____   Date: _____

Review Period: 3 years
Expiry Date: _____

1. INTRODUCTION

This procedure describes a systematic method of reviewing a computer system to:

1. Identify threats to the proper performance of the system.
2. Provide controls to mitigate these threats.

This review process is called the threats and controls review. It identifies potential quality, safety, and operability problems using a review checklist (see Section 3.3 of this procedure). The checklist ensures that all areas are covered.

2. SCOPE

This procedure applies to:

1. Threats and controls reviews of automated systems, where referenced by a validation plan.

3. PROCEDURE

3.1 Planning

The review should be performed by a team consisting of a leader and team members with expertise in the disciplines relevant to the automated system under review. The size of the review team depends on the complexity of the automated system but should consist of a minimum of two people.

The review should be performed as soon as is practicable during the validation life cycle to minimize the impact of any modifications to the requirements and specifications. This is particularly important when an external vendor is involved in the development of the automated system.

Depending on the scope and complexity of the automated system, the threats and controls review may consist of several team meetings, held during different phases of the validation life cycle. This is determined by considering:

1. The system purpose.
2. The system boundaries.
3. Which checklists (see Section 3.3) are to be used.
4. The prerequisites required.

Organizational factors such as review team membership, the number of meetings, location, and timing should be determined by the review team leader.

3.2 Review Team Meetings

Each review team meeting should be formal and minuted using the checklist defined during the planning phase above. Observations that require corrective action should be recorded on a review report and progressed in accordance with this procedure.

The review should identify potential threats that might interfere with the proper performance of the automated system. Threats can arise from the incorrect operation of hardware, software, or procedures. The review team should consider each of these when completing the checklist, using the questions as guides to focus discussion on threats to particular parts of the automated system.

For each threat identified, controls should be put in place. There are two types of controls:

1. System-related
2. Procedural

System-related controls should be tested as part of the validation life cycle. Those controls identified as procedural should be covered by written procedures that are reviewed and approved by management. Each control required should be clearly recorded on the review report.[1]

3.3 Review Checklist

The checklist should provide questions and points to consider when specifying, modifying, operating, and maintaining an automated system. The detailed checklist depends on the type of system being reviewed, which may be either a process control system or a business database system. The checklist should be agreed by management before proceeding with the review.

For a process control system, the following factors should be considered:

1. External influences (e.g., electromagnetic or radio frequency interference)
2. Quality of suppliers and services
3. Security, access and links to other systems, and electronic signatures
4. Interaction with hard-wired systems
5. Software alarms and interlocks
6. Plant and process interface
7. Human interface
8. System design and operation
9. System hardware
10. Operating system software
11. Application software
12. Maintenance and change control

For a business database system, the following factors should be considered:

1. Control of primary documents defining materials, formulated product and package components (e.g., batch sheets, package instructions)
2. Coding, and changes to coding, of materials, formulated product, and package components (e.g., lot number, part number, efficacy dates)
3. Changes of lot numbers during the process chain
4. Reports used for product release
5. Product recall reports
6. Product history
7. Stock information
8. Stock status
9. Stock location
10. Shelf life
11. Retest dates
12. Preferred suppliers
13. Tracing stock back to the supplier
14. Part use of a component or excipient
15. Split batches in production
16. Reconciliation
17. Use of electronic signature

7

Case Histories and Anecdotes

Kieran Sides

CONTENTS

I. Introduction

As well as a few light-hearted anecdotes, this chapter presents a number of case histories, in which uncontrolled changes led to unfortunate conclusions. Several more examples have been added since the first edition and one or two, which had too tenuous a connection to the pharmaceutical industry, have been removed. Thankfully, most of these instances were costly only in financial terms, and not in their impact on human life (although it is not difficult to visualize similar scenarios resulting in more devastating effects). The disdain peppered throughout some of these examples is intentional — they simply should not have happened.

II. Explosion in a New Solvent Tank[1]

An explosion occurred in a solvent tank as a result of a more rapid than usual two-stage start-up. A modification to the vent seals led to the source of ignition. No process alarms warned of anything untoward, then the second start-up sequence blew the main tank. The damage included:

- Tank roof lifted off (60 psig reached inside)
- Pipework damaged
- Seal pots blown out

The source of ignition was a new electrical heater on the seal pots, added to stop the water seal from freezing. This was added later and was not identified for full change control. Only a hazard and operability study was done on this change, by operators who were all chemical engineers. They did not spot that the element on the heaters could go above autoignition temperature (AIT) if liquid level was lost. There were also pressures on operators to get the plant running quickly. The initiating event was rapid start-up. It resulted in a large gas flowrate through the process into the solvent tankage ullage, and carried solvent vapour through the tank vent. This higher flowrate of gas blew the water out of the seal pots, exposing the hot electrical element (see Figure 7.1).

If this change had been subjected to control procedures, it is very likely that the electrical engineer would have commented that the element on the heater was not covered by the certification for use of the electrical equipment in a flammable atmosphere.

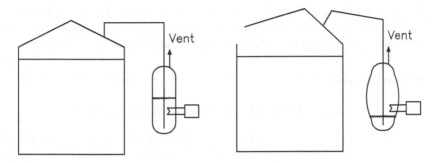

FIGURE 7.1
Schematic of seal pots arrangement, before and after explosion.

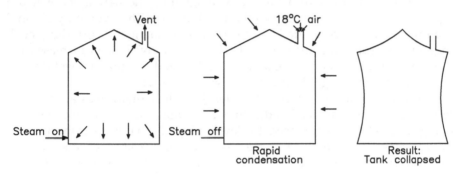

FIGURE 7.2
Tank steam out.

III. Washing, Cleaning, Sterilizing, and Purging Operations

Cleaning-in-place and sterilizing-in-place take place frequently in pharmaceutical operations, particularly where biological means of manufacture are employed. Sanders[2] describes a situation where a change of ambient temperature caused a tank failure (see Figure 7.2).

A tank was being cleaned and purged before changing to a different product campaign. The method of cleaning in this case was steam-out. However, on this occasion, the steam-out operation was done on an unusually cold day –18°C), and in consequence, the steam condensed very quickly when the steam supply was closed. Even though the atmospheric opening to the tank was more than 18 inches in diameter, the tank collapsed.

In a similar case, ammonia solution was being cleaned out from a tank.[3] It appears that water was added at a faster rate than was intended. The remaining ammonia vapours in the tank quickly dissolved in the excess water, and the tank collapsed. If ammonia is intended to be used to carry out a cleaning or sterilization operation, an increase to the rinsing rate (not usually interpreted as a potential hazard) could have similar consequences.

IV. Vessel Agitator Position Modification[3]

Although this did not occur on a pharmaceutical plant, here is a case of how batch reactors can easily run into problems with even the most seemingly innocuous minor change. A multipurpose reactor was modified to allow more cooling for one of the processes. An additional cooling coil was fitted inside the reactor. In order to allow internal space for this fixture, the vessel stirrer was replaced with a turbine-type agitator. The new agitator had a high base clearance compared to the previous one. This worked well until the process was discontinued and the (now redundant and corroding) cooling coil was subsequently removed. The removal of the coil increased the internal volume of the vessel, and hence, lowered the charge level (for the same volume of charge) in the vessel (see Figure 7.3).

During a maintenance check at a later date, the agitator was accidentally left in its topmost available adjustment position. When the simultaneous charge of reactants commenced, the level of liquid did not cover the agitator. Two reactant layers formed in the vessel as more reactants were charged. The reactor content's temperature sensor was in the lower half of the reactor. The temperature adjustment cycle adjusted the vessel temperature based on the cold, lower liquid layer temperature in the reactor. So it was actually heating at its maximum capacity instead of cooling to keep the reaction at 110°C. When the vessel contents were further increased, the agitator became covered, and the reaction started. The cooling system could not cope with the rate of cooling required for the higher rate of reaction and the higher

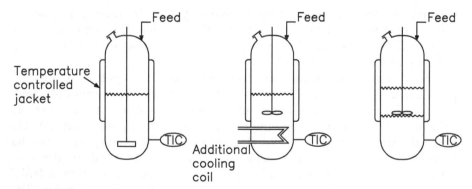

FIGURE 7.3
Vessel agitator position modification.

temperature (170 to 200°C). Other reactions occurred at the higher temperature, causing an increase in pressure, rupturing the access cover and allowing the reactor contents to overflow.

V. Room Collapse during Commissioning

There is a lot to be said for change control systems at all stages in a project, particularly if there can be knock-on effects on adjacent areas or facilities not controlled by the project.

The first phase of the project had been completed for some months and the manufacturer now boasted one of the largest cleanroom facilities in Europe. Production personnel were daily involved in the aseptic manufacture of a sterile product, each batch of which had a sales value of £1 million.

The second phase was well under way and was an extension to the first. The Building Energy Management System contractor (BEMS) for the second phase was the same company used on the first phase. Despite the protests from BEMS that their system was capable of being fully validated, it was decided to install a separate environmental monitoring system to monitor and record all critical parameters. The site incorporated a remote facility that received alarm messages from a total of 12 production facilities dotted about the site. The site was assured that this watchdog system, operated by BEMS, could only receive messages and was not capable of sending signals to change any of the parameters in the production plants, and therefore would not constitute a problem during validation.

The sophisticated controls (governing the heating, ventilation, and air conditioning systems on the project) required to maintain the high air change

The Project Manager turns away. He's seen this before.

FIGURE 7.4
Room collapse during commissioning.

rates throughout the area and the strict pressure regime had been commissioned, but the BEMS commissioning engineer attended site one weekend, merely to install the latest version of the software. There was no change intended in the operation of the system. Unfortunately, change was made.

The new version software still harbored a few bugs. The supply fans turned themselves off, and the extract fans ramped themselves up to full speed and pressure, with the consequence that the extension was sucked in. Vinyl was ripped from the walls before the walls collapsed. This resulted in many thousands of pounds worth of damage, and it was only by luck that the expensive product in the adjoining facility was not compromised.

Eventually, the extension was rebuilt at great expense to BEMS, but, of course, by this stage, the project was well behind schedule. Finally, it was nearly ready when the same thing happened again. The walls came tumbling down. But who could be at fault this time? There were no BEMS personnel on site.

A change had been made to the software by BEMS, not from the site, but from their offices 15 miles away. Validate that.

VI. Autoclave Upgrade

It was Tuesday, and lunchtime could not come around early enough. The autoclave was left to its standard sterilization cycle. It would be finished in 10 minutes, but the operator was hungry — lucky for her.

A few seconds earlier, she might not have been so lucky.

FIGURE 7.5
Autoclave upgrade.

Apparently the autoclave manufacturer's service engineer had paid them an unannounced visit that morning and, being so acquainted with the site and its working practices, knew his way around. He had many machines to upgrade, and theirs was just one of them. The quicker he was in and away, the quicker he was at the next one. After all, all he had to do was install a floppy disk and download the latest version of control software; nothing could be simpler. He would call them in the afternoon and let them know he had been there.

As she was leaving the room, the door to the autoclave suddenly swung wide open (right in the middle of its sterilization cycle), filling the room with scalding steam. It had never done that before, but then, it had never run on that version of software before.

All the outward signs said that the autoclave was the same as yesterday, and outwardly it was. If there is a way of circumventing the system, service engineers will find it, especially if the alternative means turning a one hour job into half a day.

VII. Lactose — California vs. Wisconsin

The product had been consistent for years and the tablets could not be more uniform if they were cloned. All of a sudden, the tablet machine started picking (i.e., the tablet sticking in the mould) for no apparent reason. Everything was checked, but nothing had changed. So why could they no longer produce a product conforming to specification?

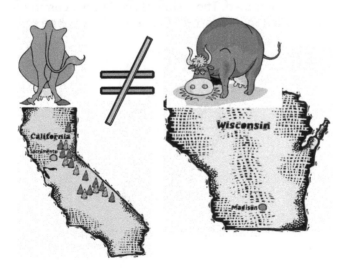

FIGURE 7.6
Lactose — California vs. Wisconsin.

Eventually, someone noticed a "C" at the end of the number on the batch of lactose, where there had previously been a "W." A bit of delving disclosed the reason why $300,000 worth of products could not be sold. The lactose supplier had neglected to inform the buyer that the company had recently been taken over. The new owner's accountants had decided things could be done more efficiently — or was it more economically? (It is amazing how many times we hear the word economy in association with the word false).

What difference could there possibly be in supplying lactose from "C"ali-fornia instead of from "W"isconsin? After all, lactose is lactose. Cows + grass = lactose. However, different cows + different grass = different lactose. Never accept the clause "the vendor reserves the right to" from a supplier. It will rarely benefit anyone but the vendor. In this particular case, it did not even benefit the vendor — he lost the contract.

VIII. Asthma Inhaler Cartridges Scrapped

It was one of the company's biggest-selling products and had become even bigger since their decision to make it available in aerosol form. They had always used the same supplier for the aerosol valves and had never experienced any problems with the quality.

One day, however, quality control (QC) checks revealed that some of the warehoused finished product was leaking. Investigating further, it was discovered that the aerosols were leaking slightly around the area where the can was crimped to the valve. All the components were checked thoroughly and the culprit was identified. The results of checks on incoming components had not been recorded in such a way that a trend was observed, but one of

Relabelling just wasn't an option.

FIGURE 7.7
Asthma inhaler cartridge scrapped.

the critical dimensions of the valves — that of the crimp seal area— had been drifting steadily toward the outer limit of its acceptable tolerance. This, coupled with the fact that the operation of the crimper was also just within specification, had been enough for the leakage to occur.

The supplier had not noticed the drift as the normal in-house quality checks had never revealed any significant difference in this particular dimension before, and complacency had crept into the checking process. Nobody had felt the need to tighten up the QC monitoring following the switch to the new change parts. After all, they had been manufactured to exactly the same drawings and specification as the last parts. The need for cost savings throughout all areas of the supplier's organization had resulted in all departments making cutbacks and this had seemed an obvious step, to move to a lower, and cheaper grade of material, while still complying with the customer's specification. The valves passed all the initial checks with flying colors, so nothing had effectively changed.

But it had, and the aerosols, which were designed to dispense 112 shots (28 days at 4 inhalations per day), now contained more like 100 shots. With nearly £1 million worth of product in the warehouse, the company toyed with the possibility of relabeling the product, with a minimum content, but eventually had to destroy the whole lot.

IX. Increase in Demand on Site Steam

A site steam system had always coped quite happily with the site's demands, so it did not seem a cause for concern when an additional autoclave was requested by the building A operations manager. A quick calculation, and the engineering manager was content.

The last time the engineering manager had been consulted about the delivery capacity of the steam system, the maximum demand in building B was achieved when the autoclave and freeze dryer were in operation and one of the holding tanks was also being sanitized. So, when the new autoclave in building A was installed, the engineering department quite rightly took the precaution of ensuring that it was operating satisfactorily with building B in its maximum demand condition.

The effects were felt across the site. Holding tank 1 could not achieve its specified heat-up rate, but eventually got up to temperature. Nobody had checked the chart recorders; some operator had just glanced at the calibrated temperature gauge and was satisfied. The product had been transferred for filling (on a filling line) before the chart records were scrutinized, and it was seen that the required holding time had been met. But it was also realized that the required temperature had only been achieved for a fraction of the specified period. The batch had to be ditched at great expense.

Elsewhere on the plant, the batch in the mixing vessel had been transferred to holding tank 2 before it was discovered that the mixing temperature had

I think we need a bigger one.

FIGURE 7.8
Increase in demand on site steam.

not been achieved for the specified time. This batch was also consigned to the drain.

Meanwhile, the autoclave and freeze dryer cycles failed, and the product undergoing treatment was scrapped. To cap it all, the water for injection system sanitization had to be repeated and a great deal of validation work had to be performed to establish the new maximum demand condition, which also meant that operating procedures in the two buildings had to be reorganized. There were also numerous other knock-on inconveniences and expenses.

X. Uncontrolled Drawing

When there are two different versions of a particular drawing, each with the same revision number, this situation can only lead to trouble.

When it was decided to create a new production area, the design engineer, responsible for process equipment cost estimating decided the quickest way to get a price for the mixing vessel was to amend a copy of the drawing for the vessel in mix room 14 and issue this with an enquiry. The mixing vessels were very similar; the main difference was the capacity (height and diameter). But, as is so often the case, the engineer had no sooner changed the sizes on the drawing when the project was put on hold. He had a feeling

The last one went in that way.

FIGURE 7.9
Uncontrolled drawing.

that the project would eventually go ahead, so he kept the partially revised drawing.

A few months later, while he was on holiday, it was decided, as a matter of urgency, to increase the output from mix room 14. Amongst other things, this involved the procurement of another mixing vessel, identical to the existing one. In the absence of the design engineer, a helpful junior obtained a copy of the drawing of the mixing vessel in mix room 14.

It is not hard to guess what happened. He took the copy that had been partially revised. He satisfied himself that the drawing number, title and content were correct and issued it for tender. He obviously had not checked every detail on the drawing, thinking there should have been no reason to. (This constituted two errors: poor document practice with the drawing, and poor engineering practice, by not ensuring a competent check of the design details before approving it for purchase.)

The design engineer returned from holiday and heard about the alterations to mix room 14. He was happy that things had progressed well in his absence and busied himself with other aspects of the project. The price was submitted, and an order was placed. There was no need to check on the manufacturer's work because that manufacturer had been used by them so many times before and had always performed admirably (poor engineering practice again). Not until the vessel was delivered, some 10 weeks later, did the problem manifest itself. The vessel did not fit in the building — it was too tall and too wide.

The vessel could have been cut down to the required size and capacity, but the characteristics of the vessel had completely changed. The capacity of the jacket for steam or cooling water was different, so the heat-up and cool-down rates would be different from the existing vessel. This would have meant operating both vessels to different procedures and running a completely different validation exercise for the new vessel. The vessel was scrapped and a new one ordered. As you would expect, the extra cost and delay to the project program were not appreciated by the business.

XI. Plant Labeling Inconsistency

Everybody always thought that tank 1 was the one on the left. All the drawings showed it in that position and all the standard operating procedures (SOPs) reflected this. The fact that the tank nameplate identified it in small, almost illegible print as tank 2 had escaped everybody's attention. Well, not quite.

A fitter had been recently seconded from another area on the site to help with workload. It was the annual plant shutdown and there was much to be done. His first task was to change the sprayball in tank 2. He was not informed of the purpose of this, only that, like everything else, it had to be done quickly.

The reason for the change was that tank 2, which was identical in all aspects to tank 1 (apart from the sprayball and nameplate), was now going to be used in the same process as tank 1. All the cleaning validation work associated with tank 1 had been carried out with a 180° upward throw sprayball (i.e., the water and steam entering the tank via the sprayball were thrown up to the lid of the tank), cleaning it and all the various penetrations, including the one for the sprayball itself. The cleaning validation, which had been time-consuming and expensive, would not all have to be repeated for tank 2, if it had the same sprayball as tank 1. A much abbreviated exercise would suffice.

The fitter duly collected the sprayball from the stores and installed it in tank 2, first making quite sure that he was at the right place by checking the tank nameplate. It did not matter to him that the new sprayball he had just fitted was exactly the same as the one he had removed. There had obviously been something wrong with it. He threw the old one away and got on with his next job.

It was some time later that quality control raised the alarm because of significant differences in the cleaning results from the two tanks. There was no reason to suspect the sprayball because all the paperwork showed that the required type of sprayball had recently been installed in tank 2, so the investigation work was centred elsewhere. Nobody could understand the results. Two or three batches of product were lost before the fault was identified. The 180° downward throw sprayball that had always been installed in tank 2 (tank 1, according to the nameplate) was still installed in tank 2. Tanks 1 and 2 were subsequently identified with two-foot high letters on the side, in agreement with the documentation. The numerical identifications on the nameplates were duly obliterated.

XII. Cheaper Cleaning Materials

Using a cheaper cleaning agent seemed an obvious thing to do; the cost savings over the course of a year would be more than considerable. Everybody had to play their part in cutting costs, and the purchasing department was given all sorts of incentives to find cheaper, but just as acceptable, alternatives.

One detergent is much the same as another, isn't it? The purchasing manager found one matching the specification almost verbatim, except that the new one would go further because it was more concentrated and could be diluted more. How was the he supposed to know about things like cleaning validation? As far as he was aware, detergents cleaned; they would not be using them otherwise. Nobody in production even noticed the difference. Why should they? The containers were the same size and color, and the labels were similar.

Analytical methods for all cleaning materials have to be validated. Because the analytical methods used in the cleaning were not the appropriate ones in this case, seven batches of an expensive product had to be destroyed. The current validation results were not applicable for the new cleaning agent and were, therefore, completely meaningless.

XIII. Batch Records Inaccurate

A SOP was changed, as so often happens, while the normal operator was on holiday and the stand-in's holiday was due to start the day after he came back. The stand-in was instructed in the new SOP, and, because there would be a day's overlap between the normal operator returning to work and the stand-in going away, the handover would include the necessary update. An air traffic controller's strike, normal at this time of year, ensured that the planned overlap did not occur.

On his eventual return, the normal operator carried on where he had left off, unaware of the change in procedure and batch records. He had no reason to read the SOP hanging on the wall. He had written it and knew it like the back of his hand. A new revision number does not exactly jump off the page and slap you across the face. So, he carried on as before. The way he was finally made aware of the change in process was extremely embarrassing and very costly for the company.

The visit by the U.S. Food and Drug Administration (FDA) investigator had been expected and dreaded for some time. Inspectors have an uncanny knack for locating flaws, and of all the SOPs the company had, she decided to focus on this one. She followed each operation, comparing it with the version hanging on the wall.

When he ended the cycle, after 3 minutes and 30 seconds, she asked him why (what a silly question; she was the one reading the SOP). In answer, he said, "that is the way it has been done for the last 4 years; it tells you at the top of page 2." There had been no such indication at the top of page 2 for the last 8 weeks. The quality assurance manager must have wanted the earth to swallow him up, but only after he had a chance to chastise the operator. Eight weeks' worth of batch records were wrong.

On the strength of what she found, the FDA investigator looked at a lot more than she had originally intended to, and in an awful lot more detail. It turned out to be an extremely expensive lesson.

XIV. Gardener's Hut

All the drawings, SOPs, and maintenance records referred to the vaccine production building as area 47. The SOPs had been revised about four times since the area had been redesignated area 12. Nobody could remember why the production areas had been renumbered, but it had to do with accounting or asset registers, or something like that.

FIGURE 7.10
Gardener's hut.

An aerial photograph hung proudly in the boardroom next to a layout drawing of the whole site. The drawing had been completed only four months ago and was extravagantly framed. It was, probably the only document on site that reflected the change in area numbers. Everyone was so used to the old area designations that nobody had adopted the new ones. So it came as a great surprise to the team fronting an FDA investigation when one of the visitors, who had been reading a vaccine production area SOP, walked across to the drawing and then asked why vaccines were being produced in the gardener's hut.

XV. The Effect of Process Tweaks over Time

The multinational manufacturing organization had produced this particular active pharmaceutical ingredient (API) on site for 15 years and had never had a bad batch. Even so, the visiting regulatory authority inspector had asked to see their validation of the process. Fifteen years ago, APIs, or bulk pharmaceutical chemicals as they were then known, were not high on the agenda for inspectors, and validation was certainly not a priority for most manufacturers. There was nothing else for it, they would have to retrospectively validate the process. What a completely pointless prospect; they knew they were well in control, because a history of batch records told them so. This was going to be no more than an expensive paperwork exercise, and the company would get nothing more out of it than the appeasement of the inspector. There was no available resource internally, so a specialist validation contractor was sought. The duly, and reluctantly, appointed contractor commenced work and found its designated points of contact less than enthusiastic in assisting with its search for existing documentation. After all, they could have done it themselves, given the time, and they would not be paying a fortune to an outside organization to tell them what they already knew.

The contractor needed to be able to identify the critical process parameters in order to meaningfully qualify the operation of the process equipment, and was not having much luck. Its quest for evidence was resulting in conflicting information. There were discrepancies between the batch record sheet, the manufacturing SOPs, the process manufacturing instructions (an overview of the process parameters, set points, and tolerances), and the drug master file. Further discrepancies came to light when the actual manufacturing operations were observed. Some of the documentation was out of date, but so what? They would change it. But what needed to be changed, and what did it need to be changed to?

The validation contractor convinced the company to let it formally document a rationale for the process. The outcome was a complete story of the manufacturing process from receipt of raw materials to dispatch of the finished API. Every parameter that could possibly have an adverse effect on the quality of the final product was identified. In an attempt to finalize and approve this document, the appropriate representatives from the production, quality, technical, and regulatory affairs departments were assembled. All the parameters were identified, and after some heated debate, those critical to product quality were agreed. Then it was time to agree the set points and tolerances for the critical parameters. This took a lot longer, but eventually the final document was approved by all concerned.

The grim reality of the last 15 years was revealed. Their process was not as robust as they thought. Of the 19 critical process parameters agreed, only 11 were acknowledged in the batch record sheet and other documents perused earlier, none of which were consistent in their recognition of these parameters. The set points and tolerances differed throughout. When the actual manufacturing process was compared to that registered, there was no contradicting the evidence; it was a different process. Fifteen years of tweaking had seen to this. A minor adjustment to the temperature here, a slight increase in pressure there, a negligible reduction in the charge rate, an insignificant reduction in agitation time, a trifling slowing of the homogenizer, an unimportant extension to hold time; all had impacted the registered process, and who knows how many times? But they were all so trivial that there was no point in recording them.

When the registered process was reestablished, the end result was not one they recognized. It was certainly not one that conformed to any of the QC specifications. Six months later, their inability to manufacture product, meeting the specification for the registered process, had cost the company millions, and there was still no sign of impending good fortune. All because there was no change control system in place to ensure that the need for the first minor alteration was reported, investigated, and approved before the change was implemented.

XVI. Climb in Product Yield

Not all uncontrolled change has a totally negative impact.

Something had obviously changed. The fermentation process normally produced a yield of 60 to 65%; all of a sudden it was 84%. Fingers were pointed at the routine maintenance the fermenter had just undergone, but it had been precisely that, routine. Nothing had been replaced, no enhancements had been made. When calibrated, none of the instruments had required adjustment. What could it be? Demand for the product was such that the process was now in operation 24 hours a day, 7 days a week, and an increase in yield to 100% would only just satisfy orders. The maintenance time had only just been afforded, so there was no way a thorough investigation could be permitted. Samples were taken and analyzed. More samples were taken and analyzed. All produced the expected results.

Four weeks of head scratching had led to the breaking of all production records, yet nothing revealed the cause. There was head scratching in the

The ladder to success?

FIGURE 7.11
Climb in product yield.

maintenance department as well; what had he done with that ladder, where could it be? He had not seen it since … no, surely not?

The head of production was not thrilled to be told an aluminium ladder may have been left inside his fermenter, but at the end of this batch, they would have a look. Sure enough, it was barely recognizable as a ladder, but there was certainly the wreckage of something wrapped around the agitator. When freed, it was evident that bits of the ladder were missing.

The cost incurred by recalling the batches of product dispatched during the last 4 weeks was painful to bear, but a bit of research into alternative methods of agitation led to a routine yield of more than 90%, so it was not all bad news. Incidentally, the new agitator was installed under strict change control.

XVII. Change to Emergency Exit Could Have Been Disastrous

Those famous words, "a lady has rung in to ask if there is going to be a hurricane tonight… there is not," will haunt Michael Fish forever. The hurricane hit Britain a few hours after the TV weatherman's assurance. All over England, there were scenes of devastation where trees had been laid low by winds gusting to in excess of 90 miles per hour.

Tree surgeons had never been busier and the pharmaceutical company's maintenance manager considered himself lucky to be able to secure the company's attendance as soon as he did. Four trees had been felled at the rear of the tablet-packing area, and some major surgery was needed to several more.

When they got around the back of the building, they had to negotiate their way through a narrow gap between the building and what was left of a once majestic oak. In doing so, one of the ladders on the roof of the van dislodged the "emergency exit — do not obstruct" sign from a door. Being a responsible individual, the driver phoned the maintenance manager, told him what had happened, and put the sign on the passenger seat. They carried on to the far end and gradually worked their way back. When he went to check on their progress, the maintenance manager found their van parked half way along the wall, not far from a different door (with no sign on it). He shouted something to them, but their ear defenders and the chainsaw rendered his words ineffective. He saw the sign in the van, assumed incorrectly, temporarily fixed the sign to the door and decided to come back later. Having cleared the way, they moved the van further along the wall, this time parking with the back end partially obstructing a door (the earlier one — now with no sign on it). It all looked different coming from the other direction and they had forgotten about the sign.

What happened next was not a life-threatening incident but it could have been. One of a number of storm-related problems had resulted in a short circuit in the adjacent plant room and the appearance of smoke in the packing area. The operators were not in any immediate danger, but the section leader ordered an evacuation of the area. They calmly headed for the emergency exit as the smoke became more visible. When the door was opened, the operators realized they could not fit through the partially obstructed opening, and mere concern led to a certain amount of anxiety. They could not see the van but could see fallen branches through the gap. They presumed a branch was in the way and a couple of the men put their shoulders to the task, but two and a half tons of van were staying put. Vivid imaginations began working overtime and thoughts turned to those horrific incidents we hear about on television where fire investigations reveal charred corpses piled up against locked fire exits. Anxiety did not stop at trepidation but carried on straight through to panic. When one person started screaming, several others joined in — but they could not match the din of the chainsaw.

The plantroom door opened, and an engineer explained that the short had caused some wiring to burn out in one of the control panels, but it had been contained, and what was all the noise about? So, Michael Fish is not the only one who would like to forget that time. The unplanned change to the emergency exit route had left the packing department with one dislocated shoulder, a good deal of mental anguish, and a large helping of embarrassment.

XVIII. Water for Injection System Contamination

In 6 years the water for injection system had never let them down. Apart from the occasional failure attributed to sampler error, the QC laboratory had never seen a sample remotely threatening the specification limit. Why then, all of a sudden, were they starting to see microbiological contamination? Speculation abounded. Maybe a biofilm had built up to such a level on the pipework internals that some of it had become dislodged and was now being circulated around the system. Extensive sampling throughout the system revealed no obvious source of the problem. The system was drained, refilled, and a sanitization cycle was performed. All was well again — though not for long.

Water for injection is expensive stuff to produce and now, 6 months later, they had to ditch thousands of litres on a regular basis. But why? The company operated a strict change control system, and a review of the recent changes had shed no light on the predicament. It was not until each of the maintenance fitters was interrogated that the only possible culprit was identified. One of the zero deadleg valves at one of the sample points on the system had been damaged externally and had been replaced by another identical unit during the last shutdown period, when all the valve diaphragms and system seals had been replaced. It had not come to light when auditing the change control system because the fitter had assumed it was a like-for-like change, which the system unfortunately allowed without the normal documented justification. But the shutdown had been several weeks before the first sample failure. If a problem had been caused by this work, it would surely have surfaced earlier.

Despite the fact that there was no reason for the replacement to have made a difference to the water quality, the maintenance manager, who was now clutching at straws and was satisfied nothing else had changed, decided it had to be investigated. The new valve was indeed identical to its predecessor — same manufacturer, same model number, same size. There was no arguing, it was a like-for-like change. The valve, which was held in place by Triclover™ fittings, was removed for further inspection. Although the valve seemed fine, it appeared that the upstream seal had been pinched on installation and was protruding into the pipe. But how would this account for the introduction of microbiological contamination? The seal was replaced.

In the quality control laboratory, they were also investigating, and it appeared there was a trend to the contaminations. Because the company had such a good history of water quality, it had been able to reduce its level of routine sampling to only a couple of points at any one time. Sample points were rotated so that samples were taken from each point on different days of the week and then the cycle was repeated. The out of specification results always occurred following the sampling from this particular valve. This had to be more than coincidence. The valve manufacturer was consulted and a computerized model of the installation was generated. The flow rate was calculated and the pinched seal was added. There it was! The encroachment of the seal into the pipe was causing a venturi effect, and as the sampler opened the valve to take a sample, a small amount of air was being drawn into the system. The resulting microbiological contamination to the system was always witnessed the next time the system was sampled.

Greater care was now exercised when changing seals, and the system returned to being the robust performer it had previously been. So, the replacement valve was not actually the cause of the problem, but it is unlikely the investigation would have led to the valve being removed, and the rogue seal spotted, quite so early, had the damage to the existing valve not occurred and had the seal been replaced as part of a routine change.

XIX. Room Clearance Failure

A change to personnel roles meant that it was unclear precisely whose job it was to verify the room was cleared between production campaigns, but everybody assumed it had been done. The equipment had been cleaned and all was set for the next product. The active ingredients were added and manufacture commenced. Unfortunately, the product was supposed to have only one active ingredient, the one left from the previous run should not have been there. Still, as far as the operator was concerned, it looked the same and it would not have been there if it was not meant to go in, would it?

The in-process quality checks picked it up almost immediately, so there was no threat of harm to a patient, but two extremely expensive active ingredients were wasted. It is not just the obvious physical amendments to a process that need to be controlled. Less apparent changes to operations can also have knock-on effects that must be recognized and countered.

XX. Drain Valve Opened by Mistake

Being the school holiday period, they were short staffed, and a couple of operators were hired in from the agency. They had worked on site at least once before so the minimum amount of effort would have to be expended in bringing them up to scratch. A quick run through the plant and 15 minutes with the SOPs should be enough. After all, it was not the most complicated of processes. They were confident they had taken it all in and were ready to start.

Now, what did they say about the end of the distillation stage? Where is that SOP? Ah, yes, open valve 19 until all the product has transferred. Now, where is valve 19? Great, there it is. Just verify it from the tag. This is child's play.

Valve 61 was duly opened and 5000 litres of product were slowly transferred, straight to drain. It's no use blaming the operator; he did not know there was a valve 61. It did not get a mention in the SOP he had been given. And if 19 and 61 are indistinguishable, it is an accident waiting to happen.

Personnel changes have to be properly managed. Skimping on training will always lead to expensive mistakes.

XXI. Increase in Compressed Air Capacity

The company was going to be installing three new packing lines, the operation of which would require significant amounts of compressed air. It was an ideal time to make the changes discussed for some time. The oiled compressors would be pensioned off and three new oil-free compressors would be installed. The capacity of the system would be increased by approximately 40%.

There had been, and still were, times when simultaneous demands on the system meant insufficient pressure or flow at certain use points. Such occurrences need never happen again. Even though the existing compressors were not oil-free, testing at several representative use points dotted across the site had demonstrated that particulate contamination of the air was negligible, the pressure dew point was never going to result in a moisture problem, and the presence of hydrocarbon could not be detected. This evidence was the only reason the regulatory authority had let them continue to manufacture.

The change notice was accepted because all the changes appeared to be positive. Capacity would be increased, quality would be improved, the inspectorate would relax. The existing distribution system would have to stay. It had grown arms and legs over the years and now consisted of several miles of pipework, comprising all manner of sizes and materials, but replacing it would bring production on the site to a halt and the thought could not even be entertained.

Once installed the new compressors certainly produced the goods. Suddenly, demand was satisfied at all use points on a continuous basis. There was only one problem; the increase in pressures and flow rates ensured that the dirt, rust, and hydrocarbons that had settled quite happily at the bottom of the pipe over the years could now be uplifted to become unwanted ingredients in products site-wide.

The cost of contaminated product was considerable. When added to the extra cost of filtration at use points, it was more than considerable.

XXII. Thalidomide[4]

In 1958, Chemie Grünenthal launched and commenced selling a new brand-name sedative, Contergan. In the U.K., it was marketed as Distaval by Distillers Limited. It is known universally as thalidomide. The drug was widely available in more than 40 countries and was available without prescription as a treatment for morning sickness. Just as the application for marketing approval in the U.S. was made in 1960, a correlation between mothers who had taken Thalidomide and babies born afflicted with phocomelia (literally, "seal limbs") was identified, first, in Australia, and quickly followed by supporting data from Germany. (As it turned out the S-thalidomide isomer was responsible for the deformities. The R-thalidomide isomer has been shown to be fine.)

The publicity and substantial financial settlements obtained by the afflicted point out the risks associated with launching all new drugs. This incident was a landmark case since thalidomide was the first drug to be implicated in widespread deformities in embryos. It exposed weaknesses in medical knowledge and was catalytic in initiating how drug testing should be carried out for teratogenic effects.

XXIII. Southern Corn Leaf Blight[5]

In 1970, southern corn leaf blight damaged about 15% of the U.S. corn crop. Seed corn producers responded by making unusual efforts to produce blight-resistant varieties of corn by genetic engineering. In 1971, the U.S. Department of Agriculture encouraged farmers to plant much larger crops of corn than previously — about 20% more. These changes, when combined, resulted in a large corn surplus and a sharp fall in corn prices. Farmers had two consecutive bad years, one by losing a significant proportion of their crop, and one by overproduction. These changes focused on the problems associated not only with monoculture farming but also changes made as a result of genetic engineering, without adequate research and planning.

XXIV. The Great Cranberry Scare[5]

In 1959, U.S. cranberry growers applied a promising new herbicide (amitrole) in a manner that had not yet been approved. This lack of change control was their nemesis because the following changes were taking place about the same time:

- Legislative changes that required the listing of chemicals found to be carcinogenic in laboratory tests were underway.
- The FDA set no limit of acceptability for amitrole at the time.
- Improved analytical techniques that could determine previously undetectable levels of the herbicide on cranberries were being developed.
- Government advised consumers not to buy cranberries as a precaution, in response to the laboratory results.
- Thanksgiving was approaching.
- There was massive publicity about the potential hazards.

As one might expect, these events led to public confusion and a negative impression, apparently far beyond the actual risk posed, and hence, a huge fall in demand for cranberries. It took until 1964 for the cranberry industry to recover.

XXV. The Cyclamate Affair[5]

Cyclamates, a group of nonnutritive sweeteners, had been approved by the FDA for use as a food and drink sweetener in 1950. However, in 1969, the U.S. government announced that cyclamates had been found to cause cancer in rats in long-term feeding studies. Therefore, no cyclamates were to be sold after January 1970. The announcement had a severe effect on the companies producing cyclamates and using them in their products.

Some companies had been aggressively selling their cyclamate-containing products without complete knowledge of their effects. The key change here was the new test results, carried out on behalf of the U.S. government, coupled with the introduction of new legislation prohibiting carcinogenic substances in foods. The tests ought to have been undertaken by the manufacturers, who also should have assessed the potential impact of changes in legislation on their business.

XXVI. Anecdotes and Quotes

"That's Not My Job"

This is the story about four people named Everybody, Somebody, Anybody, and Nobody. There was an important job to be done and Everybody was sure that Somebody would do it. Anybody could have done it, but Nobody did it. Somebody got angry about that, because it was Everybody's job.

Everybody thought Anybody could do it, but Nobody realized that Everybody would not do it. It ended up that Everybody blamed Somebody when Nobody did what Anybody could have.

Development of Murphy's First Law

The more innocuous a design change appears, the further its influence will extend.[6]

Picric Acid (trinitrophenol) Reaction[7]

There was a young chemist from Ealing
With trinitrophenol was dealing
But, he added red lead
And, the truth must be said
They found him a splash on the ceiling

Obsolescence[6]

How simple life once used to be,
So nonchalant, and so care-free.
We dwelt on the complacent side,
Then heard about thalidomide.
Of many hazards science spoke:
Carcinogens and cigarette smoke,
Grave warnings about DDT,
And tuna fish and mercury.
While wondering what else was at stake,
We parted with sweet cyclamate.
The phosphate perils they came next.
Now saccharin has failed the test.
Of each new banning that we learn,
There is one utmost concern,
How will the human race stay clean —
Without hexachlorophene?

References

1. Turner, S.G. and N. Vaughn, Solvent tank explosion during commissioning, IChemE, Record of Case Study Seminar Presentations, Rugby, U.K., 1997.
2. Sanders, R.E., *Management of Change in Chemical Plants — Learning from Case Histories*, Butterworth-Heinemann, Oxford, 1993.

3. ESCIS, *Thermal Process Safety, Expert Commission for Safety in the Swiss Chemical Industry*, Booklet 8, Case 4, Basle, Switzerland, 1993.

4. Elmsley, J., *The Consumer's Good Chemical Guide*, Corgi, Reading, U.K., 1994.

5. Lawless, E.W., *Technology and Social Shock*, Rutgers University Press, New Brunswick, NJ, 1977.

6. Bond, J., *The Hazards of Life and All That: A Look At Some Accidents and Safety Curiosities, Past and Present*, Institute of Physics Publishing Ltd., London, 1996.

7. Weber, R.L., *Science with a Smile*, Institute of Physics Publishing Ltd., London, 1992.

8

Projects

Simon G. Turner

CONTENTS

I. Introduction

Chapter 5 gives a general model for controlling change. This chapter applies the general model to projects, specifically, bringing together the author's own experiences of working in a premier contracting organization. Some Foster Wheeler energy forms have been included to illustrate how one might formalize selected logging and tracking functions. I must, however, emphasize that this chapter is not Foster Wheeler's project change control procedure or an abridgement of it. Foster Wheeler's procedure is far more comprehensive than the suggested model for projects and is also fully integrated into the Foster Wheeler quality management system. This chapter is a simple proposal, to assist you with generating your own project change control protocol. You should be able to develop your own project change control procedures from information contained in this chapter.

You may find terminologies used in a contracting environment different to a manufacturing environment, but the principles for controlling change are the same. This chapter uses typical contracting project management language, so be aware of the variations in terminology from that used in Chapter 5.

In the process industries, the typical major project formula involves employing a process engineering and construction contractor. A project usually comprises a business concept, studies or preliminary design, bidding stage, contractor selection and contract award, front-end design, detailed engineering and design, procurement, construction, commissioning, and handover to operations (Figure 8.1). There is likely to be some negotiation of terms of business before contract award. This is only a selection of combinations of work scope definitions that could make up a process industries project. Each party that has a major interest in a project will nominate a project manager to represent its interests for the life of the project. Perhaps 90% of what a project manager deals with is associated with managing the process of change during the life of the project.

II. What to Control on Projects

Projects have several things in common: they represent a major investment for any operator, they are scoped, they are timed, they have to meet predefined technical criteria, and they have a budget allocated as a resource.

A. Project Attributes

The best list of the key attributes of a project that I have come across is found in *Project Management for the Process Industries*.[1] The list is summarized as

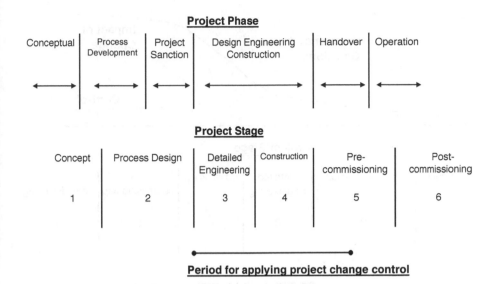

FIGURE 8.1
Project phases and project stages.

follows: contents are unique, the result of a project is to achieve something, completion of activities to satisfy customer criteria, start and end times, and bringing together appropriate resources. This reference is well worth regular consultation by any reader working in a project management role, and it can also help those newer to working in a project execution environment to understand the principles outlined in this chapter.

B. Attributes of Successful Projects

Project Management for the Process Industries also describes the following key attributes of *successful* projects:

- The customer is clearly defined.
- Customer requirements are clearly understood and agreed upon (i.e., what is to be achieved?).
- Scope, cost estimate, schedule and any critical dates are clearly predefined and agreed upon by all associated parties. (The agreement encourages commitment to meeting targets).
- Risks and uncertainties are understood and accepted by all associated parties.
- Resource allocations and resource delivery times are defined.
- Objectives, priorities and responsibilities are clearly defined for all associated parties.
- The project team is committed.

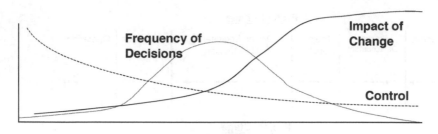

Project Stage

Concept	Process Design	Detailed Engineering	Construction	Pre-commissioning	Pre-commissioning
1	2	3	4	5	6

Period for applying project change control

FIGURE 8.2
Degree of influence and impact of change through project stages.

- Communications and the flow of project data are accurate, appropriately distributed and correctly timed.
- Safety, quality, cost and progress are monitored and controlled throughout the life of the project.
- Of course, changes are controlled effectively.

In order to understand change control on projects, you must have at least a foundation understanding of the above definitions and attributes. The relative impact of change and ability to influence change vary as a project progresses (Figure 8.2). By reviewing the impact of changes that affect the success of a project, one should be able to achieve effective change control and avoid compromising success criteria; that is the major part of a project manager's job. This is also part of the role for any key project team member; indeed every project team member.

As previously discussed in the introductory pages, we control change by applying procedures. Projects are no exception.

III. Project Change Control Procedure — What Should It Cover?

The procedure should address the three main concepts described in Chapter 5: (1) identify the change, (2) evaluate effects of the change, and

(3) authorize as appropriate and record all relevant data about the change. The procedure should outline:

- Definition of terminology used to avoid any confusion or misinterpretation
- Who is responsible for implementing the requirements for addressing the detail of the main concepts so that everybody knows their role in controlling project change
- Methodology and standard forms to facilitate: change identification or flagging, change evaluation, change authorization and recording

It is the responsibility of the project (i.e., project manger) to ensure that the project change control procedure is suitable and efficacious at controlling change. Each project must prepare its own procedure, customized to the project, parties involved and the peculiarities of the particular project.

A. Definitions

Include all project abbreviations and a glossary, to avoid misunderstandings. (For example, the difference between change flags, change requests and change orders.)

Change flag — an identified change, usually the result of discussion between customer or client and contactor engineering discipline; equivalent to engineering change request in Chapter 5 (an example of a form that Foster Wheeler uses on projects for flagging changes is shown in Figure 8.3)

Change request — a written customer or client initiated change

Change order —a reviewed and approved change flag or change request

B. Responsibilities

The procedure should at least define who is responsible for the following:

- Overall responsibility for change control
- Change flag review
- Change order approval
- Review of change impacts
- Estimating the impact (cost, schedule, etc.) of each change and expediting the information required to complete estimates (usually involves expediting returns from disciplines to ascertain impacts of change orders across the project)
- Keeping a log of all change returns, as an audit trail for the impacts of each change

FW

CHANGE FLAG

No.

Originator...

DATE.....................

CLIENT:
PROJECT:
CONTRACT NO.:
Brief Change Details:

GROUP LEADER..   DATE.......................................

PROJ COORD...   DATE.....................
PEM (AFC Docs)..   DATE.....................

INTERNAL CHANGE ☐   CONTRACT CHANGE ☐   REJECT ☐

IMPLEMENT CHANGE ☐   AWAIT INSTRUCTION ☐

Order of Magnitude Estimate of Change:
MHS..
SCHEDULE...
Material Cost...   Contract Cost.................................

Other Groups affected:	PROC	PE	PIPE	CIV	MECH	VESS	TECH
INST	ELECT	DESSFT	QA	HO CON	SITE	PM	ARCH
Others................................							

Discipline Group Leader (yellow)
Project Coordinator   (blue)
Change Coordinator   (pink)

FWEL 4144 Rev.0

FIGURE 8.3
Example of a change flag form.

- Carrying out the role of project change coordinator
- Alerting the project manager to changes that have significant implications for the project
- Identifying (i.e., flagging) changes to the project change controller; this is usually everybody on the project team, but could be filtered through lead discipline engineers to streamline submissions
- Issuing relevant written project instructions as a result of approved changes
- Maintaining change logs and summaries
- Change coordination and implementation of change orders, within agreed budget and schedule for the change
- Instill, educate, promote, and maintain a change awareness philosophy on the project
- Coordinating regular change reviews

C. Methodology

The best way to represent a systematized method for any management function is usually a flowchart, similar to those used in Chapter 5, for implementing the three types of change in a manufacturing environment. A typical project change control flow chart is shown in Figure 8.4.

1. Change Flag Generation

All team members should have a brief outline of how to fill in a change flag form, including how to number it (if the change coordinator wants them to), and what to do about ensuring the relevant log is completed. The timing for submission and to whom it is to be sent to for review need to be clearly stated in the procedure.

2. Special Considerations for Split Site Working

Ensure that other offices involved in the project, especially the construction site office, have clear instructions for special circumstances appropriate to their location. The project should treat the separate sites exactly the same as the project home office so that no unidentified changes slip through the net.

3. Feedback to Change Flag Originator — Approval Notification

After review and approval or rejection, the change flag originator should be advised promptly. This is so the originator knows what has happened to the flag they raised. The originator can then implement the change, if it has project approval, or it can be filed for no further action. All disciplines should receive regular reports on change control, which should include an up-to-date change

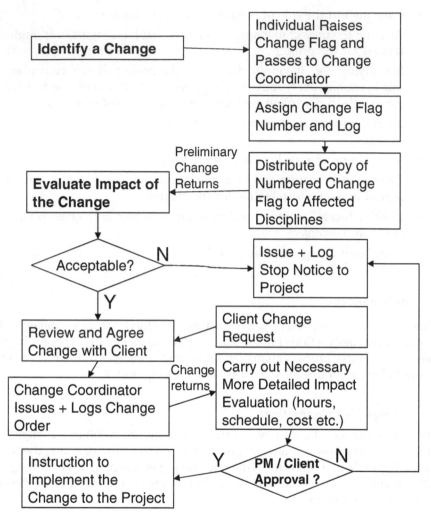

FIGURE 8.4
Typical project change control flow diagram.

log. There is nothing to be gained from keeping discipline leaders in the dark; it is simply good management to keep as many people on the project fully informed about how project changes are being controlled and implemented.

4. Change Review and Evaluation

Once a change flag form is received by the change control coordinator, a brief review with the project manager (or nominee) will determine whether more time should be spent determining the impact on the project. Depending on the terms of the contract (lump sum or reimbursable) the project manager might be bound to discuss every change flag with the client. The project manager may reject some changes outright on the grounds that they would not be compatible with the contract or success criteria for the project. The other change flags would need a more complete evaluation of impacts on the project in order for a decision to be made whether to approve them for implementation. The way to do this is to generate a change order notice, distribute it around the project, requesting impacts on each discipline; noting that man-hours, schedule, and cost impacts should be returned to the change control coordinator in a given timescale. The change order notice could be a mark up of the change flag, but my experience indicates that it is better to use a dedicated change order notice form. When requesting information from the project disciplines, a separate change returns form could be used, but a change order notice with the discipline name can suffice. An example of a change order notice is shown in Figure 8.5.

Once all the change returns for a given change order come to the coordinator, the coordinator can consolidate the estimated impacts for that change on a change summary form (Figure 8.6). If a lot of changes are likely on the project, the consolidated impacts for each change could be recorded in a summary document, a change returns log (Figure 8.7). A change returns log would support any audit trail and can also be used to compare actual impacts of the implemented change with the estimates from change returns. This is also a good way to benchmark change returns for accuracy and completeness, by direct comparison with change returns logs from other projects that had similar changes. If the project manager felt particularly concerned about change return accuracy and completeness and regarded returns as critical to the management of change, it might be beneficial to have a basis for estimating impacts outlined in the procedure.

Regular change reviews, to review change flags and change returns, should be held. Decisions about whether or not changes are approved for implementation are made at such meetings. Separate change review meetings should be held — a regular one for change flag screening and a regular one for those changes that need review and approval by the client. The nominated change control coordinator is the ideal person to coordinate these meetings. All the relevant paperwork can be brought along for review and approval (or rejection). The implementation status for each change can be reported back to the project manager. Any delays with implementing changes can be targeted for expediting here. The results of the reviews should be logged at each review meeting and the status of log entries made available to the project.

D. Standard Forms

Appropriate forms for identifying, estimating cost impacts, estimating schedule impacts of changes made in the project design office, and estimating schedule impacts of changes made at site during the construction phases are as follows.

Change flag form (Figure 8.3)

Change order notice (Figure 8.5)

Change log showing combined change flag and change order record (Figure 8.7)

Change Order Notice		
Project: _____		**Change Flag No.** _____
Project No. _____		**Change No.** _____
Location: _____		**Revision:** _____
		Date: _____

Description of Change

Reason for Change

Order of Magnitude Estimate

	Man-hours	**Costs**
Estimate Preparation		
Design Office Engineering		
Construction Site		
Subtotal of Contractor's Services	_____	_____
Equipment & Material Costs		
Construction Costs		
Escalation & Contingency		
TOTAL	_____	_____

Schedule Impacts

Design Office

On Project Completion Date

Approvals

tick as appropriate

Change Approved	☐
Prepare Cost Estimate Only	☐
Special Instructions	☐ _____
Contractor's Project Manager	_____
Client Nominee/s	_____

FIGURE 8.5

Example of a change order notice. (From Foster Wheeler Energy Limited Quality Net. Standard form 4144. With permission.)

	CONTRACT CHANGE SUMMARY	CHANGE No:	
FOSTER WHEELER ENERGY LIMITED	XYZ PROJECT CONTRACT NUMBER 1234	CHANGE FLAG :	
		REVISION	DATE

COST CODE	DISCIPLINE	HOURS TO ESTABLISH CHANGE	HOURS TO CARRY OUT CHANGE	TOTAL HOURS FOR CHANGE	ADDITIONAL COSTS
8100	PROCESS				
8231	PIPING ENGINEERS				
8232	PIPING DESIGNERS				
8311	CIVILS				
8331	ARCHITECTS				
8332	ARCHITECTS TECHNICIAN				
8341	BS ENGINEERS				
8341	BS DESIGNERS				
8410/70	MECHANICAL				
8430	VESSELS				
8440	TECHNOLGY				
8480	HEAT EXCHANGERS				
8531	ELECTRICAL ENGINEERS				
8532	ELECTRICAL DESIGNERS				
8551	INSTRUMENT ENGINEERS				
3552	INSTRUMENT DESIGNERS				
8560	DESIGN SAFETY				
8820/21	PROJECT ENGINEERING				
8600	MAT MAN				
8711	HO CONSTRUCTION				
			TOTALS		

EFFECT ON SCHEDULE

REMARKS

FIGURE 8.6

Example of a contract change summary form. (From Foster Wheeler Energy Limited Quality Net. With permission.)

FIGURE 8.7

Change Log Showing Combined Change Flag and Change Order Record

Change Flag No.	Date Raised	Originator	Subject	Discipline	Submitted to Change Controller	Passed for Client Review	Change Order No.	Man-hours Impact	Costs Impact £	Schedule Impact	PM and Client Approved	Implemented? / Comment
1	6-May-03	AN Other	Dryer design	Process	Y	N						
2	22-June-03	J. Smith	Reactor size increase	Process	Y	Y	1	200	5000	+6 weeks	Y	11/21/03
3	23-June-03	A. Jones	Pump Motor configuration	Mech	Y	Y	2	50	−2000	None	Y	
4	3-August-03	J. Bloggs	Revised centrifuge location	Piping	Y	Y	3	600	+10000	+6 months	N	Rejected
			Totals					850	5000	—	—	—

E. Approvals

It is usual for all changes to be approved by both contractor and client project managers. As indicated earlier, this is dependent on the terms of the contract. If the change impacts construction, it should also be reviewed and approved by the construction manager. If the construction manager has sole responsibility for site work, that manager could be the sole approver of those changes that only affect construction. All approval decisions should be available to the project team.

F. Change Logging and Monitoring of Approvals

A project should keep a live log of all its uniquely numbered change flags and all uniquely numbered change orders. These logs should summarize the dated status of each flag and change order, and the associated cost of each change. The change order log is usually the only means the project can keep a running tally of increasing cost of changes. See the change log showing combined change flag and change order record (Figure 8.7).

IV. Auditing

Remember that auditing how the project change control procedure is implemented will show up any weaknesses in how the project's team members are implementing the procedure. This will increase the confidence of senior managers that the management of change is under control. It will also give the project manager peace of mind that there are unlikely to be surprises later in the project (when it is very difficult to respond to them) resulting from changes.

V. Typical Pitfalls in Project Change Control

A. Instructions Coming Directly from Customer and Client Organization Are Implemented without Review and Approval by Project Management

This can be very common where the customer is regarded as very important to a contractor's business or the client project team has an autocratic style. Discipline engineers are unwilling to question the client's instructions — and implement them — without considering how instructions could be a change to scope, quality, or cost and how they could potentially significantly impact the project's success parameters. The solution is to educate project team members as to the key success attributes of the project early on in the project.

B. Poor Understanding of Project Success Criteria

In particular, lack of knowledge about scope and schedule at engineering discipline level can lead to serious problems in identifying changes. As before, the solution is to educate project team members as to the key success attributes of the project early on in the project.

C. Holding on to a Change Flag for Too Long before Deciding What To Do about It

The change might be a very necessary one, but if decision making is left too long, the normal course of design may have progressed such that the change requires already completed work to be undone. To avoid unnecessary increase in project costs, written project instructions can be issued to hold the scope associated with the relevant change flag. Project progress monitoring can be used to determine the latest time that a decision can be made, on scope that is on hold.

D. Disparate Team Members without Day-to-Day Contact with the Project Manager or Project Task Force

This situation is becoming far more common because manufacturing organizations have rationalized their design capability. Therefore, the client team is very thinly spread, perhaps across eight or nine projects, executed in several countries. It is frequently the case that contact between the manufacturing team and the home office task force design team will be almost exclusively by e-mail, supplemented by telephone discussions, which are even less controllable than e-mail. Agreements about the design are made in writing, but critically, these agreements can bypass the change control procedure and could compromise the project's success criteria. All team members need to be particularly on guard to ensure changes are flagged when elements of the project team operate in this manner. With disparate team configurations, the project manager will probably not be able to do enough to prevent all uncontrolled elements slipping through, and regular reminders to the project team should reduce the likelihood of it happening. Design quality audits (not in the scope of this book) should identify any design changes that actually do slip through the change control net.

E. Losing Track of Technical Developments to Engineering Deliverables in between Formal Issues

One might find two disciplines developing their design along different lines, because they do not have consistent up-to-date information on a piping and instrument diagram. One way to overcome this is to issue forms that outline the relevant design change and deliverables affected. This would be done outside the scope of the project change control procedure because it covers

	CONTRACT ENGINEERING DATA SHEET			CED No	/	/

FW Contract No: Sheet 1 of _____

Client: Date: _____

Project name: Originating Group: _____

Subject: Originator: _____

PART 1 – TO BE COMPLETED BY ORIGINATOR FOR ALL CED'S

Part 1a REASON FOR CHANGE	Part 1b	ASSESSMENT
	CODE	REASON
	A	- Client Request
	B	- New Supplier Data
	C	- Rev'd Supplier Data
	D	- Removal of Hold
	E	- Eng. Correction
	F	- Eng. Development
	G	- HAZOP
	H	- Other

O	Part 1c	CONTRACT CHANGE?	Yes/No	Part 1d	MATERIAL CHANGE	Yes/No
R	Part 1e	WAS CHANGE FLAG RAISED?	Yes/No			
I	Part 1f	CHANGE DESCRIPTION				

Projects	Fired Heaters
Process Eng	Design Safety
Piping	Construction
Technology	Planning
Instruments	Estimating
Electrical	Mat Man
Heat Ex	Sub-Con
Civil	Cost Eng
Vessels	Chg Co-ord
Mechanical	

Part 1h   DRAWINGS AFFECTED

PART 2 – TO BE COMPLETED BY CED CO-ORDINATOR / PROJECT ENGINEER

Part 2a – DETAILS OF IMPACT ON MATERIALS / EQUIPMENT / SERVICES
Ordered Urgently / Ordered as part of next MTO*
Reduced / cancelled for existing orders*   (*delete as required)
Part 2b – IS REVIEW BY DESIGN SAFETY REQUIRED?   Yes/No

PART 3 – AUTHORIZATION FOR THE AMENDMENT TO DATABASE
PROJECT ENGINEERING MANAGER CHANGE APPROVAL

SIGNED _____   DATE _____

DISTRIBUTION: Mark up Distribution as required

Projects	Instruments	Vessels	Construction	Estimating
Process Eng.	Electrical	Mechanical	Design Safety	Mat Man
Piping	Heat Exchangers	Chg Co-ord	Fired Htrs	Planning
Technology	Civils	Sub-Con	Cost Eng	

FWEL1706 Rev 4

FIGURE 8.8
Example of contract engineering data sheet form (for interim deliverable developments).

interim issue change, not changes that impact the project. At Foster Wheeler, these changes are recorded by the discipline that originates the deliverable change on contract engineering data sheets (CEDs) (Figure 8.8). The approval is by the project manager or nominee. Remember that changes like these are not intended to be changes relevant to the contract, they are part of normal design development. These changes are only likely to threaten the success of the project if they are not technically reviewed and approved, then properly advertised to all relevant team members.

Reference

1. Lawson, G., Wearne, S., and Isles-Smith, P., *Project Management for the Process Industries*, The Institution of Chemical Engineers, 1999.

List of Abbreviations

AADA	Abbreviated Antibiotic Drug Application
AIChemE	American Institute of Chemical Engineers
ANDA	Abbreviated New Drug Application
API	Active Pharmaceutical Ingredients
BACPAC	Bulk Actives Chemical Postapproval Changes
BFS	Blow Fill Seal
BIOS	Basic Input Output System
BLA	Biologics Licensing Application
BPI	Bulk Pharmaceutical Ingredient
BSE	Bovine Spongiform Encephalopathy
CBE	Changes Being Effected (number of days to implement usually indicated, e.g., CBE-30)
CBER	Center for Biologics Evaluation and Research (U.S.)
CDER	Center for Drugs Evaluation and Research (U.S.)
cGMP	current Good Manufacturing Practice (ongoing development of GMP)
CHAZOP	Computer HAZOP
CIP	Clean-in-Place
CJD	Creutzfeld-Jakob Disease
CMC	Chemistry, Manufacturing and Controls
COSHH	U.K. Control of Substances Hazardous to Health Regulations
CVM	Center for Veterinary Medicine (U.S.)
DI	Design Improvement
DIC	Design Improvement Change (same as DI)
DMF	Drug Master File
DQ	Design Qualification
ECR	Engineering Change Request
EDQM	European Directorate for the Quality of Medicines
EEA	Engineering Equivalence Assessment

ELA	Established License Application
ELD	Engineering Line Diagram
EPA	U.S. Environmental Protection Agency
EU	European Union
EWR	Engineering Work Request
FDA	U.S. Food and Drug Administration
FDAMA	Food and Drug Modernization Act
FD&C Act	Federal Food, Drug, and Cosmetic Act (U.S.)
FMEA	Failure Modes and Effects Analysis
GA	General Arrangement Diagram
GDP	Good Document Practice
GEP	Good Engineering Practice
GLP	Good Laboratory Practice
GMO	Genetically Modified Organism
GMP	Good Manufacturing Practice
HVAC	Heating, Ventilation and Air Conditioning
HAZAN	Hazard Analysis
HAZOP	Hazard and Operability Study
ICH	International Conference on Harmonization of Technical Requirements for Registration of Pharmaceuticals for Human Use
IChemE	Institution of Chemical Engineers
IMPs	Investigational Medicinal Products
IND	Investigational New Drug Application
IQ	Installation Qualification
IR	Immediate Release Dosage Form
IVRS	Interactive Voice Recognition System
MA	Marketing Authorization
MAF	Master File for a Device (or Device Master File)
MCA	U.K. Department of Health Medicines Control Agency
MHRA	Medicines & Healthcare products Regulatory Agency
ML	Manufacturer's License
MRA	Mutual Recognition Agreement
MSDS	Material Safety Data Sheet
NDA	New Drug Application
OQ	Operational Qualification
PAC-SAS	Postapproval Changes to Sterile Aqueous Solutions

PAC-ATLS	Postapproval Changes to Analytical Testing Laboratory Sites
PAI	Pre-Approval Inspection
PAS	Prior Approval Supplement
P&ID	Piping and Instrument Diagram
PDA	Personal Digital Assistant
PDUFA	Prescription Drug User Fee Act
PHS Act	Public Health Service Act (U.S.)
PICS	Pharmaceutical Inspection Cooperation Scheme
PLC	Programmable Logic Controller
PQ	Process Qualification
ptfe	Polytetrafluoroethylene
QA	Quality Assurance
QC	Quality Control
QP	Qualified Person
RA	Regulatory Affairs
SARS	Severe Acute Respiratory Syndrome
SHE	Safety, Health and Environment
SIP	Sterilize-in-Place
SMF	Site Master File
SOP	Standard Operating Procedure
SQ	System Qualification
SUPAC	Scale-Up and Postapproval Changes
SUPAC-IR	Scale-Up and Postapproval Changes for Immediate Release Dosage Forms
SUPAC-MR	Scale-Up and Postapproval Changes for Modified Release Dosage Forms
SUPAC-SS	Scale-Up and Postapproval Changes for Non-sterile Semisolid Dosage Forms
SUPAC-TDS	Scale-Up and Postapproval Changes for Transdermal Systems
TSE	Transmissible Spongiform Encephalopathy
vCJD	Variant Creutzfeld-Jakob Disease (BSE in man)
WFI	Water for Injection
WQ	Works Qualification

Glossary

Albumin Any group of simple water soluble proteins that are coagulated by heat and are found in blood plasma, egg whites, etc. (also Albumen).

Change control The principle of change control is to provide a formal mechanism for identifying and reporting all changes to the facility, equipment, raw materials, processes, procedures, or any other activity that could have an impact on the final product or business. The aim is to provide an auditable trail for managing any change.

Change flag An identified change written on a form, intended for project review. Usually arising as a result of discussion between customer or client and contactor engineering discipline on a project. equivalent to engineering change request in a manufacturing environment.

Change levels SUPAC Guidance value, 1 to 3, to indicate the likelihood that product quality or efficacy could be adversely affected by a change; 3 being most likely.

Change order A reviewed and approved change flag or change request on a project.

Change request A written customer- or client-initiated change.

Change request log Document or database that is dedicated to recording the engineering change process. Provides information on the current status of an engineering change request (ECR) and the people assigned to the change. Tracks progress through the approval stages and gives details of the location of the final documentation.

Change return Submission of consequential impacts of making a change. These impacts are requested of disciplines, on a project, by a change control coordinator and returned by the disciplines to the change control coordinator. Such returns are usually the person-hour impacts of the change, but would include any relevant purchasing cost increases and schedule impacts.

Consequence The result of an event or action. The ultimate consequence may be different from the immediate consequence.

Contract facilities Those establishments contracted by the firm to provide services and activities in the design, development, production, testing, or distribution of finished products (i.e., blenders, sterilizers, test laboratories, packagers, etc.). These facilities are considered by the FDA to be an extension of the manufacturing facility.

Design improvement   Small, often low-cost modification to equipment or system to bring about greater efficiency, ease of use or process improvement. These are life cycle events, but also occur during the design stage of a larger project as designs mature or as a result of issues in commissioning (same as minor modification).

Design review   Process of systematically assessing a design to ensure compliance with the user's requirements, company standards and good engineering practice. In some companies, this might be within the hazard study system.

Ebola   A severe and often fatal viral disease causing one form of African haemorrhagic fever.

Equivalence   Substituting one item with another that has similar performance but not direct like-for-like replacement. Typically, this is a life cycle issue of obsolescence on an existing plant, where the original part is no longer available.

Event   An incident that causes a system or system element to move from one state to another.

Excipient   An inert substance that is combined with an active drug for preparing a convenient dosage form (e.g., sugar or gum).

Filing   A document detailing the processes, equipment, and procedures used to manufacture a regulated product and approved by the regulatory authority.

Frequency   The likelihood of an event, always expressed as a rate (e.g., number of occurrences per year, number of occurrences per second). Not to be confused with probability.

GENHAZ   A system for the critical appraisal of proposals to release genetically modified organisms (GMOs) into the environment.

Good document practice   Ensuring that key documents are created, reviewed, approved, distributed, and stored in a controlled manner.

Good engineering practice   Execution of engineering activities according to a formal quality management system that ensures proper handling of requirements, design, review, and audit of deliverables throughout the engineering life cycle.

Good manufacturing practice   (Refer to EC Orange Guide 1992 and U.S. Food and Drug Code of Federal Regulations, title 21).

Hazard   Anything that has a potential to damage a system or its environment. May be a condition, a physical or chemical characteristic, a human action, a natural element, a software bug, etc.

Hazard review   Hazard study.

Hazard study A methodical sequence of activities to identify, evaluate, assess, control, and document the hazards and risks in a system throughout the complete life cycle of the system. Often includes ergonomics and operability assessments.

Iatrogenic Caused by process of diagnosis or treatment.

Investigational new drug application Protocol in the U.S. that allows an unapproved drug to be shipped between states to facilitate clinical studies.

Leukocytes (or Leucocytes) White or colourless corpuscle of blood found in lymph, etc.

Life cycle An orderly pattern of behavior over a period of time characterized by a set of states and events.

Likelihood Frequency or probability.

Minor modification Seemingly trivial changes, the consequences of which can easily be grossly underestimated and lead to major problems. Typically those inexpensive jobs that appear to need no approval (see also design improvement).

Minor project Modification to equipment or systems requiring purchase of components, interruption to manufacturing or operation, and generally involving some level of financial authorization.

Modification A change to a system, normally intended to improve the system in one or more aspects. This kind of change can usually be identified if it is referred to by the following keywords: replace 'a' by 'b,' alter, add, change, remove, or improve.

Oligonucleotides A small section of a DNA molecule, usually up to about ten base pairs long.

Peptide A molecular structure term used to describe a molecule made up of a string of amino acid molecules.

Pfiesteria A red blooming algae that causes bleeding and death of fish.

Picking An effect, potentially affecting tablets, where the tablets become stuck in their mold during manufacture.

Probability The likelihood of an event. Always expressed as a dimensionless number — usually a fraction of 1 — and occasionally as a percentage. Not to be confused with frequency.

Process A sequence of activities or transformations performed to achieve a defined goal.

System A system is a combination of many parts that work together toward a common goal. A system will consist of many subsystems and will exist within a larger system, usually called its environment.

Threat Business term for hazard: Anything that has a potential to damage a system or its environment. May be a condition, a financial or contractual characteristic, a human action, a natural element, a software bug, etc.

User requirement specification   User defines the key attributes of a system or component for use by the designer in producing the Functional Specification. The URS is cross-referenced in the Operation and Performance Qualification stages of validation to ensure compliance with the original design intent.

Validation   The action of proving that any material, process, procedure, activity, system, equipment, or mechanism used in manufacture or control can, will, and does achieve the desired and intended result(s). This is a key concept of GMP.

Validation master plan   An overall plan that, when executed, will produce documented evidence that the production facility is validated.

Zero dead leg valve   A valve designed to be installed in a branch so as to present no static hold up of fluid between the main branch and the valve when in the closed position.

Index

Milton Keynes UK
Ingram Content Group UK Ltd.
UKHW040056071024
449327UK00019B/601